Football Math

Touchdown Activities and Projects for Grades 4–8

Second Edition

Jack Coffland and David A. Coffland

A
GOOD
YEAR
BOOK™

A
GOOD
YEAR
BOOK™

Good Year Books are available for most
basic curriculum subjects plus many
enrichment areas. For more Good Year
Books, contact your local bookseller
or educational dealer. For a complete
catalog with information about other
Good Year Books, please contact:

Good Year Books
P.O. Box 91858
Tucson, AZ 85752
www.goodyearbooks.com

Illustrations by Doug Klauba.
Book design by Patricia Lenihan-Barbee.
Additional pages designed by
Performance Design.

Introduction for Parents and Teachers

From *Football Math: Touchdown Activities and Projects for Grades 4–8* published by Good Year Books. Copyright © 1995, 2005 Jack Coffland and David A. Coffland.

When giving students in grades 4 through 8 problems to solve, we must be certain that they have practiced a wide variety of problems. By the time students are finished with the eighth grade, they should be proficient with problem-solving involving:

- Whole Number Computation
- Fraction Computation
- Decimal Computation
- Percent Computation

One problem classification system used by many mathematicians includes both routine and non-routine problem situations, as described below. In other words, it is no longer appropriate to give students problems that simply review computational operations that have just been taught.

Routine problems

"Routine problems" are defined as those problems that ask students to apply a mathematical process they have learned in class in a real-life, problem-solving situation. This book defines two types of routine problems:

1. Algorithmic problems—These are word problems (story problems) that ask students to read the problem, figure out the computational procedure required, and then apply that computational algorithm to solve the problem. For example:

Bobby scored 6 touchdowns in last night's game. Each touchdown is worth 6 points. How many points did he score in all?

2. Multi-step problems—These are algorithmic problems that demand two or more computational steps in order to obtain the answer. For example:

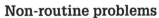

Last year the Bruisers won three-quarters of their 12 games. This year they won 5 games. How many wins did they have over the two seasons?

Non-routine problems

In recent years math educators have focused additional energy on "non-routine" problems—those that challenge the learner in some way. The different kinds of "non-routine" problems in this book are challenge problems and mini-project problems.

1. Challenge problems—Problems of this type are non-routine in that the student does not know how to solve them from memory. They require the use of heuristics, the act of inventing steps. It is the true test of problem-solving ability. Examples are:

A one-kilometer train enters a one-kilometer tunnel moving at 30 kilometers per hour. How much time will pass from the time the engine enters the tunnel until the caboose comes out of the tunnel?

Problems of this type are the final challenge in math. We cannot quit until we have challenged students to invent or create solutions to problems. The professional scientist, engineer, or mathematician all work to create ideas, not to simply rehash old ideas. But the myth of mathematics learning has always been that only people in these professions must solve problems. The truth of the situation is that every day the carpenter, the clerical worker, or the grocery store clerk also invent solutions to problems.

2. Mini-project problems—These problems are "process" problems, not simple story problems. They are often open-ended in that different students may obtain different answers. The process is more important than the product; the process stresses such things as multiple steps, differences in answers, and discussion of considerations to see if everyone agrees. For example:

How much money do we need to take with us on the field trip to the stadium?

Notice that this situation depends upon several different variables; not everyone will take the same amount of money on the trip. Solving mini-project problems teaches children that not all problems have simple answers, nor do they all have one answer.

Long-term projects

Finally, because this book is meant to capture the interest of students by combining mathematics and football, we have suggested project problems. These are not really math problems; they are projects that the student can undertake that require the use of math and a knowledge of football. They are meant to be fun and to make math and football the student's hobbies.

Resources

If students are interested in learning more about football and its mathematical possibilities, take a look at these resources:

- Your local paper will print statistics of local high school or college teams.

- The National Federation of State High School Associations prints (or publishes on the Internet) high school football rule books, statistic books, and other materials.

- The NCAA publishes both books and Internet sites on college records.

- The NFL also publishes books and/ or Internet sites on NFL football records.

- *Sports Illustrated* magazines often discuss records; they also publish Web sites and books on sports records.

- Go to ESPN web sites for extensive football information, from histories to record charts.

The advantages of looking on the Internet first is that such records are either up-to-date or the site will include a note explaining that the records are through the 2004 season.

From *Football Math: Touchdown Activities and Projects for Grades 4–8* published by Good Year Books. Copyright © 1995, 2005 Jack Coffland and David A. Coffland.

This book is about football; it contains a great deal of interesting material about football—professional football, college football, and even high school football. But it is also about math. It asks you to solve math problems that stem from football statistics, stories, and situations.

This material attempts to explain some interesting things about football. For example, you can see how a college passing rating is determined. You will be given a problem to figure out one example yourself, but you will enjoy the book much more if you tackle the project on rating college passers. Collect statistics on your favorite quarterback; then see if you can figure out his rating before you read it in the paper. Or, if you are playing football, keep track of your own statistics and rate yourself!

The book also contains a number of facts about college and professional football. For example, who holds the career rushing record for the Dallas Cowboys? What college team had the best record during the last 10 years? The information is presented in math problems—have fun solving them or give them to your friends to solve. You will already know the answers. Enjoy!

Contents

From *Football Math: Touchdown Activities and Projects for Grades 4–8* published by Good Year Books. Copyright © 1995, 2005 Jack Coffland and David A. Coffland.

Contents

Contents

From *Football Math: Touchdown Activities and Projects for Grades 4–8* published by Good Year Books. Copyright © 1995, 2005 Jack Coffland and David A. Coffland.

Contents

Activities

Miami Dolphin Records

Solve the following problems.

1. In 1984 Dan Marino set many of the NFL's single season passing records. He completed passes for 5,084 yards in 16 games that year—a record. How many yards did he average for each game?

_____

2. In 1984 Dan Marino set the Dolphin record for throwing 48 touchdown passes in one season. Because each touchdown is worth 6 points, how many points did the Dolphins score on Marino passes that year?

_____

3. Dan Marino's favorite targets for many years were Mark Clayton and Mark Duper, known as the "Marks Brothers." Clayton caught 550 of Marino's passes, while Duper caught 480. How many passes did they catch all together?

_____

4. Several years later the Dolphins gained 1,525 yards rushing and only 3,975 yards passing. A team's total offense is calculated by combining the rushing and passing yards. What was the Dolphins' "Total Offense" figure for 1992?

_____

5. During the same 1992 season, the San Francisco 49ers led the league in total offense with 6,369 yards gained. How many more yards did the 49ers gain than the Dolphins?

_____

6. Larry Csonka was the fullback on Miami's undefeated team in 1972, and he still holds the career rushing record for the Dolphins. He was also involved in one of the NFL's strangest plays. During a game, Csonka received a penalty for unnecessary roughness as he carried the ball and the opponents were trying to tackle him! His stiff-arm was "too forceful!" Csonka's career rushing statistics for the Dolphins include 1,506 carries for a total of 6,737 yards. What was his "average yards per carry" for the Dolphins?

_____

Silver and Black Attack

Many famous players have spent some time with the Oakland (or Los Angeles) Raiders. The team often signed older players who had been cut by other teams, and they won ball games. Here are stories of some of those players. Most did not play their entire career for Oakland; you might want to look them up (in record books or on the Internet) to see the entire history of each of these players.

1. George Blanda was a quarterback/kicker who played in 348 games in his career. He entered the NFL with the Chicago Bears in 1949 and ended his career with Oakland. Jim Marshall, who played on the defensive line with Cleveland and Minnesota, was second on the list of games played. He appeared in 282 games. How many more games did George Blanda play?

_____

2. Tim Brown holds the Oakland record for passes caught (1,070) and for touchdowns (102). (This was Brown's record at the end of the 2003 season; he did enter the 2004 season, so he will be adding to his records.) How many points did Oakland score on Tim Brown's touchdowns? (Remember that he did not make the extra points!)

_____

3. Jerry Rice played for the San Francisco 49ers for many years; he is considered by most to be the greatest wide receiver ever to play the game. But as he approached 40 years of age, he signed to play with Oakland. He holds the NFL record for "most passes caught." At the end of the 2003 season, Rice had caught 1,519 passes. How many more passes has Rice caught than Tim Brown?

_____

4. Marcus Allen holds the career rushing record for the Raiders, but he played all of his years as a Los Angeles Raider, not an Oakland Raider. He gained 8,545 yards as a Raider. He also played for Kansas City, so he gained a total of 12,243 yards rushing in his entire career. How many yards did he gain while he played for Kansas City?

_____

Challenge problems

1. Jim Marshall (mentioned in the first problem) is remembered for one play when he was with Minnesota. Do some research to find what that play was.

2. Morton Anderson is a kicker who has played for several teams. He came back to the NFL for the 2004 season, when he was 44 years old. Has he broken George Blanda's record for most games played?

Debbie's Dogs

Debbie and her club have decided to run the concession stand at the high school football game as a way to raise money for the club. She wants to figure how much food she should buy and how much they sell for each game. You can help her.

1. Debbie knows that the high school football field will seat 3,000 people. The principal has been able to predict that, based on past years, half of those people will buy a hot dog. How many hot dogs should Debbie buy if she expects a sell-out crowd?

_____

2. Debbie also knows that about two-fifths of a crowd will buy coffee. How many cups should she buy if the athletic department expects to have a sell-out?

_____

3. For the first game, the crowd is small. Only 1,434 people attend the first game. If two-thirds of them bought a cold drink (it was a hot night), how many soft drinks did Debbie and her club sell?

_____

4. During the second game, Debbie decided to try selling a new snack—nachos. The attendance at the game was 2,464 people, and Debbie sold nachos to three-quarters of them. How many orders of nachos did Debbie sell?

_____

5. For the game played on Halloween night, Debbie and her crew dressed in their best costumes and served hot cider. The crowd was small because most people stayed home to meet the "trick or treaters." Only 1,566 people came to the game, but five-sixths of them had the hot cider. How many cups of hot cider did Debbie sell that night?

_____

6. During the last game, it was really cold, but 2,985 people attended the game anyway. Debbie counted the coffee cups that were used and figured that four-fifths of the people bought a cup of coffee. How many cups of coffee did Debbie and her club sell that night?

_____

From *Football Math: Touchdown Activities and Projects for Grades 4–8* published by Good Year Books. Copyright © 1995, 2005 Jack Coffland and David A. Coffland.

Debbie's Dollars and Sense

The club members worked so hard selling things that they wanted to know how much money they made. You can help Debbie figure the profits from last night's game.

1. Soft drinks make more money for the club than anything else they sell. Soft drinks come in only one size; it sells for a $1.00. From that dollar, 82¢ is profit. After counting cups, Debbie knows that she sold 934 soft drinks. How much money did the club make on soft drink sales?

_____

2. Hot dogs sell for $1.50. Debbie and her club make 78¢ on each hot dog sale. If they sold 576 hot dogs last night, how much money did they make on hot dog sales?

_____

3. Candy bars are not a good money maker. They cost $1.00, but only 36¢ of that is profit. If they sold 164 candy bars last night, how much did they make selling candy bars?

_____

4. A cup of coffee sells for only 75¢, but 55¢ of that is profit. How much did Debbie and her hard-working crew make selling coffee last night if they sold 634 cups of coffee?

_____

5. Nachos sell for $2.00, and $1.05 of that is profit. Debbie sold 1,246 containers of nachos and cheese last night. How much profit did the club earn on those sales?

_____

6. Figure the total profit Debbie and her club made last night on the items listed in problems 1 through 5.

_____

Challenge problem

Do you think that taking over the concession stand was a good way for Debbie's club to make money? You may need to know that they worked six home games, and they made about the same amount of money for the other nights as they did on this night.

From Football Math: Touchdown Activities and Projects for Grades 4–8 published by Good Year Books. Copyright © 1995, 2005 Jack Coffland and David A. Coffland.

Great Teams and Great Players

From 1982 through 1995, San Francisco played in five Super Bowls and won all five games. (The only other undefeated Super Bowl teams have played and won their only Super Bowl game.) The chart shows team Super Bowl won/loss records for teams with five or more Super Bowl appearances.

Records of Teams with Five Super Bowls (through 2004)

Team	Total Games	Total Wins	Total Losses	Winning Percentage
San Francisco 49ers	5	5	0	1.000
Pittsburgh Steelers	5	4	1	0.800
Dallas Cowboys	8	5	3	0.625
Oakland/L.A. Raiders	5	3	2	0.600
Washington Redskins	5	3	2	0.600
Miami Dolphins	5	2	3	0.400
Denver Broncos	6	2	4	0.333

1. Joe Montana, quarterback for four of San Francisco's five Super Bowl wins, holds the career record for most TD passes thrown. He completed 11 passes for touchdowns. How many points were scored on Montana's TD passes?

2. Montana also holds the career record for most pass completions in Super Bowls. He completed 83 passes in his four Super Bowl appearances. How many completed passes did he average in his four Super Bowl games?

3. Jerry Rice, who holds most pass reception records (most passes caught, most yards gained passing, most touchdowns scored, etc.), also holds the record for most TDs score in Super Bowls. He caught eight passes for touchdowns in his five games. How many points did Rice score in Super Bowl games?

4. Roger Craig was the 49ers' running back in several Super Bowls. He also holds the most 49er rushing records. He had some of his best days against the Denver Broncos. He gained 417 yards rushing in just four games against the Broncos. On the other hand, he gained only 72 yards in three games against the Pittsburgh Steelers. How much better was his average rushing mark against the Broncos than against the Steelers?

5. Steve Young was the quarterback for San Francisco's 1995 Super Bowl victory. In 1992, he led all NFL quarterbacks in the efficiency rating statistics. He ran for 537 yards and passed for 3,465 yards. Because he was injured briefly, assume that he played in 13 complete games. What was his combined average for both rushing and passing for each of those games?

Deee-fense

Make the following computations about the Frontier League defensive players and answer the questions that follow.

Frontier League Defensive Statistics

Defensive Player and Position	Total Sacks	Tackles Made	Fumbles Recovered	Passes Intercepted
Larry Minor, LB	2	42	6	1
Jon Davisdon, DL	14	23	1	0
Stan Smith, LB	12	67	4	1
Jon Echohawk, CB	1	24	2	9
Howard Gardner, DL	6	28	1	0
Claude Gibbons, CB	0	56	2	3

Position Code: LB=Linebacker; DL=Defensive Lineman; CB=Corner Back

1. Larry Minor recovered 40% of the fumbles that his team recovered. How many fumbles did his team recover all together?

   _____

   If Larry's team played six games this season, would you say the coach is happy with the number of fumbles his defensive team recovered? Why or why not?

   _____

   _____

   _____

   _____

2. When you total all of his defensive statistics, you find that Stan Smith was responsible for ending 84 plays. If the defense was on the field for 276 plays in their six games, what percentage of the plays did Stan end personally?

   _____

   Do you think Stan is the star of his defensive team? Why?

   _____

   _____

   _____

   _____

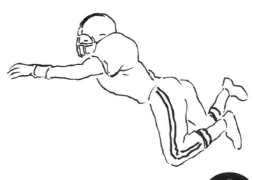

From *Football Math: Touchdown Activities and Projects for Grades 4–8* published by Good Year Books. Copyright © 1995, 2005 Jack Coffland and David A. Coffland.

Deee-fense (cont'd.)

3. Jon Echohawk intercepted nine passes; only 30 passes were thrown in his direction. What percentage of the passes thrown to players on his side of the field did Jon intercept?

_____

If you were the coach of a team that had to play Jon, would you throw many passes to his side of the field? Why or why not?

_____

4. Claude Gibbons made 40% of all the tackles made by his team. How many tackles did his team make all together?

_____

Do you think Claude plays on a team with a winning record? Why or why not?

_____

_____

_____

Beasts of the East

During the 1980s, the NFC Eastern Division was known as a tough league where teams liked to play "slam-bang" football. Almost every team had a tough defense and an offense where they tried to run the ball down each other's throats. Two of those tough 1980s coaches have now returned to the Eastern Division—Joe Gibbs to the Redskins in 2004 and Bill Parcells to the Cowboys in 2003. It will be interesting to see what happens in the NFC's East Division now.

NFC East Conference Statistics, 1980s

Team	Rushing Stats	Passing Stats
Cowboys	1,713	3,890
Eagles	1,512	3,808
Giants	1,516	4,044
Redskins	1,247	4,109

1. Look at the individual season rushing statistics. Find the average number of rushing yards for four teams. (Carry out your answer to the nearest tenth.)

2. Now look at the single-season passing statistics for each of the teams. Find the average number of passing yards for these four teams. (Again, carry your answer out to the nearest tenth.)

3. Now use your averages to answer this question: How much larger is the average passing record over the average running record for these teams? (Carry your answer out to one decimal point.)

Challenge problems

Do NFL East teams still stress running the ball?

Find the running statistics for these teams during the 2004 (or later) NFL seasons. Compare those figures to the statistics listed above. Are the NFC East teams rushing for more or less yardage?

Find the passing statistics for these teams during the 2004 (or later) NFL seasons. Compare these figures to the statistics listed above. Are the NFC East teams passing for more or less yardage?

Bobcats Victory

It was a wild ball game at the high school last night. The score was Bobcats 48, Grizzlies 42. Let's look at some of the statistics to see why the score was so high.

1. Bobby Nelson, a Bobcats wide receiver, caught 12 passes. He averaged 14.72 yards per catch. How many total yards did he gain on his pass receptions?

   _____

2. Timmy Johnson, the Grizzlies quarterback, completed 34 passes for 387 yards. What was his average gain per completion? (Figure your answer to the nearest hundredth.)

   _____

3. Big Jon Crump, the Bobcats fullback, ran for 137 yards on 16 carries. What was his average gain per carry? (Round your answer to the nearest tenth.)

   _____

4. There were only three punts in the game. The Grizzlies punted all three times for 121 yards. What was the punter's average distance per kick? (Round your answer to the nearest tenth.)

   _____

5. The Grizzlies had a hard time running the ball. Three Grizzly running backs gained only 48.7 yards, 23.2 yards, and 17.5 yards, respectively. How many total yards did the Grizzlies gain rushing the ball?

   _____

Challenge problem

When a score is this high, do you think the offensive team improved their statistics? Do you think a defensive team improved their statistics? Why or why not?

From Football Math: Touchdown Activities and Projects for Grades 4–8 published by Good Year Books. Copyright © 1995, 2005 Jack Coffland and David A. Coffland.

Why Is It Called Football?

From *Football Math: Touchdown Activities and Projects for Grades 4–8* published by Good Year Books. Copyright © 1995, 2005 Jack Coffland and David A. Coffland.

NFL defenses are so strong that a team wants to score every time they are in the "red zone." It is also important that a team comes away with at least a field goal if the offense drives inside the opponent's 40-yard line. For that reason, field goal kickers are extremely important offensive weapons. Look at the following statistics for four of the better kickers in the National Football Conference.

Field goal statistics, National Football Conference, 2003

Kicker and Team	Extra Points PATs	Field Goals	Total Points
Jeff Wikins, STL	46/46	39/42	163
Ryan Longwell, GB	51/51	23/26	120
Jason Hanson, DET	26/27	_/23	92
Martin Gramatica, TAM	33/44	16/26	92

Answer the following questions about these statistics:

1. How many total points did all four kickers score?

   _____

2. If Wilkins made 39 field goals, and if each field goal is worth 3 points, how many points did he score on field goals?

   _____

3. How many field goals did Jason Hanson make if he scored 66 of his points on field goals?

   _____

4. How many more points did Ryan Longwell score than Martin Gramatica?

   _____

5. What was the total number of extra points kicked by all of the kickers listed?

   _____

Challenge problem

After examining the statistics listed above, fill in the missing items for Carolina Panthers kicker J. Kasay:

Extra Points	Field Goals	Total Points
_/30	32/38	125

11

They Get a Kick Out of This

Use the table below to solve the following problems.

Field Goal Statistics, American Football Conference, 2003			
Kicker and Team	Extra Points PATs	Field Goals	Total Points
Mike Vanderjagt, IND	46/46	37/37	157
Jason Elam, DEN	39/39	__/31	134
Adam Vinatieri, NE	37/38	25/__	112
Orlando Mare, MIA	33/34	22/29	99

1. Jason Elam attempted 31 field goals. He made 87% of his attempts. Fill in the blank on the chart that tells how many field goals he made.

2. These four kickers missed only 2 points after touchdowns (PATs). What was the percentage of kicks made for the total of all the kicks attempted by these four kickers?

3. Adam Vinatieri of New England made 25 field goals; he made 73.5% of the kicks he attempted. Fill in the blank that tells how many field goals he attempted during the season.

4. Now that you have completed the chart, find the percentage of field goals that were made for all of the field goals attempted by these four kickers.

5. A kicker really helps his team when he makes a 50-yard or longer field goal, because his team then scores points when they did not get close to the goal line. Orlando Mare attempted six kicks from beyond 50 yards. He made four. What percentage did he make?

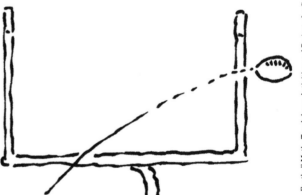

From Football Math: Touchdown Activities and Projects for Grades 4–8 published by Good Year Books. Copyright © 1995, 2005 Jack Coffland and David A. Coffland.

AFC Rushing Leaders

F or a running back in the National Football League, the goal is to rush for 1,000 yards in a season. Eleven players in the American Conference accomplished that goal during the 2003 season.

AFC Thousand-yard Rushers, 2003

Player	Team	Att	Yards	Avg.	Long	TDs
Jamal Lewis	Ravens	387	2,066	5.3	82	14
LaDainian Tomlinson	Chargers	313	1,645	5.3	73	13
Clinton Portis	Broncos	290	1,591	5.5	65	14
Fred Taylor	Jaguars	345	1,572	4.6	62	6
Priest Holmes	Chiefs	320	1,420	4.4	31	27
Ricky Williams	Dolphins	392	1,372	3.5	45	9
Travis Henry	Bills	331	1,356	4.1	64	10
Curtis Martin	Jets	323	1,308	4.0	56	2
Edgerin James	Colts	310	1,259	4.1	43	11
Eddie George	Titans	312	1,031	3.3	27	5
D. Davis	Houston	238	1,031	4.3	51	8

1. How many more yards did Jamal Lewis gain than LaDainian Tomlinson?

_____

2. Clinton Portis and Eddie George both changed teams after the 2003 season. What is the total number of yards they gain for their old teams? What is the total number of times they carried the ball for their old teams?

_____

Do you think these running backs will be missed by their old teams?

_____

3. Two of the most famous backs, Eddie George and Priest Holmes, found it difficult to break loose for long runs. What was the difference between the longest runs for each player?

_____

4. Ricky Williams had the most carries; D. Davis had the least number of plays in which he carried the ball. How many more times did Williams carry the ball than Davis?

_____

5. Priest Holmes scored 27 touchdowns for his team, the Kansas City Chiefs. How many points did he score for his team with those touchdowns?

_____

6. Jamal Lewis gained more than 2,000 yards in one season. The list of players rushing for more than 2,000 yards in a season is very small. How many more yards did Lewis gain than two very famous running backs, Eddie George and Edgerin James, combined?

_____

Challenge problem

Can you find a list of all the running backs who have gained more than 2,000 yards in one season?

_____

From *Football Math: Touchdown Activities and Projects for Grades 4–8* published by Good Year Books. Copyright © 1995, 2005 Jack Coffland and David A. Coffland.

NFC Rushing Leaders

From *Football Math: Touchdown Activities and Projects for Grades 4–8* published by Good Year Books. Copyright © 1995, 2005 Jack Coffland and David A. Coffland.

For a running back in the National Football League, the goal is to rush for 1,000 yards in a season. Seven players in the National Conference accomplished that goal during the 2003 season:

NFC Thousand-yard Rushers, 2003

Player	Team	Att	Yards	Avg.	Long	TDs
Ahman Green	Packers	355	1,883	5.3	98	15
Drew McAllister	Saints	351	1,641	4.7	76	8
Stephen Davis	Panthers	318	1,444	4.5	40	8
Shawn Alexander	Seahawks	326	1,435	4.4	55	14
Tiki Barber	Giants	278	1,216	4.4	27	2
Anthony Thomas	Bears	244	1,024	4.2	67	6
Kevan Barlow	49ers	201	1,024	5.1	78	6

1. How many yards did Ahman Green and Drew McAllister gain together?

_____

2. Shawn Alexander scored 14 touchdowns. How many points did he score for his team?

_____

3. Ahman Green carried the ball more than any of the other players, while Kevan Barlow carried it fewer times than any of the other players. How many more carries did Ahman Green have than Kevan Barlow?

_____

4. Ahman Green had the longest run from scrimmage for any of these players; Tiki Barber had the shortest "long run." How much longer was the Ahman Green's longest run than that of Tiki Barber?

_____

5. Carrying the ball more than 300 times in the NFL is difficult. Four of the NFC players who gained more than 1,000 yards carried the ball more than 300 times during the year listed. Assuming that each played in all 16 regular-season games, how many carries did they average in each game?

Ahman Green _____

Drew McAllister _____

Stephen Davis _____

Shawn Alexander _____

Challenge problem

What does the "average run" statistic tell you? Discuss what happens each time Ahman Green carries the ball. Compare that to what happens each time Anthony Thomas carries the ball.

_____

Home Field Advantage

Much is made of the "home field advantage" in sports. It always seems easier to win at home, because the home team does not have to travel. But interesting cases can be made for those teams that play in extreme conditions, such as high altitude or extreme weather. Four teams whose home game might present "extreme conditions" are the Miami Dolphins and the Tampa Bay Buccaneers, where visiting teams face extreme heat; the Denver Broncos, where visitors must play in the "mile-high city," or at a high altitude; and the Green Bay Packers, where teams may face cold or snowy conditions.

Home and Away Records, 1994–2003

Team	Home Wins	Home Losses	Road Wins	Road Losses
Miami Dolphins	56	24	40	40
Denver Broncos	60	20	38	42
Green Bay Packers	68	12	40	40
Tampa Bay Buccaneers	52	28	34	46

1. Denver's old stadium was called "Mile High Stadium" and the city is called "the mile-high city." It is difficult for teams to perform well at high altitudes, so at one time Denver had the highest home-field winning advantage of any NFL team. For the 10-year record shown in the chart, what was Denver's home field winning percentage? What was its winning percentage for road games?

   _____

   _____

2. Miami's games are usually played in the afternoon, but Miami afternoons can be hot and humid, even in December. As games go on, teams from cooler climates often seem to wilt in the intense heat. What was Miami's home field winning percentage in the years between 1994 and 2003? What was its road winning percentage?

   _____

   _____

3. Of the three teams, Green Bay had the most home wins in the 10-year period shown. Calculate the percentage of home wins for Green Bay. Then, can you state Green Bay's winning percentage for road games, without doing any calculations?

_____

4. Tampa Bay had the fewest wins of the four teams listed, but it still seemed to have a home field advantage. Calculate the home field winning percentage for Tampa Bay; and then calculate the road game winning percentage for the Bucs. How would you describe their road record for the 10-year period?

_____

5. Combine the records for all four teams to find the total number of home wins and losses and the total number of road wins and losses. Then, for all four times combined, calculate the total winning percentage for home and away games.

_____

_____

_____

_____

_____

From Football Math: Touchdown Activities and Projects for Grades 4–8 published by Good Year Books. Copyright © 1995, 2005 Jack Coffland and David A. Coffland.

All-time Winning Coaches

A t the end of the 2003 NFL season, there were seven NFL coaches with more than 200 total career wins. They were:

NFC Coaches with 200+ Total Career Wins

Coach	Years as Head Coach	Total Wins			Regular-season Wins		
		Won	Lost	Tied	Won	Lost	Tied
Don Shula	33	347	173	6	328	156	6
George Halas	40	324	151	31	318	148	31
Tom Landry	29	270	178	6	250	162	6
Curly Lambeau	33	229	134	22	226	132	22
Paul Brown	25	222	112	9	213	104	9
Chuck Noll	23	209	156	1	193	148	1
Dan Reeves	23	201	174	2	190	165	2

1. In 1993, Don "Shoes" Shula passed George "Papa Bear" Halas to become the coach who won more games than any other coach in NFL history. Shula went on to add to his record for several more years. At the end of their careers, how many more "Total Wins" did Shula have than Halas? How many more "Regular-season Wins" did he have?

_____

2. The chart shows both the total win/loss record and the regular season win/loss record for each coach. What is the number of wins and losses that each coach had during the "second season"—wins in play-off, league championship, and Super Bowl games?

_____

2. Chuck Knox, the former coach of the Rams, Bills, and Seahawks, would be eighth on the total win list, but with 193 total wins, he is 7 short of making the list of coaches who have won 200 games. As of 2004, Marty Schottenheimer is the active coach with the most wins; he has coached the Browns, Chiefs, Redskins, and Chargers to a total of 170 wins. How many games must Shottenheimer win to pass Knox on the list of "Coaches with the Most Wins"?

_____

3. Very few coaches stay with the same team for their entire career. For example, Don Shula coached for both the Baltimore (now Indianapolis) Colts and the Miami Dolphins. But George Halas, Tom Landry, and Chuck Noll accomplished all of their victories with one team. Halas only coached the Chicago Bears; Landry only coached the Dallas Cowboys; and Noll only coached the Pittsburgh Steelers. How many games did these three coaches win all together?

_____

4. Curly Lambeau coached the Green Bay Packers, the Washington Redskins, and the old Chicago Cardinals. He was a head coach in the NFL for 33 years. On the average, how many games did he win each year? (Because you can't win part of a a game, don't worry about your remainder when you finish doing the problem!)

_____

5. Dan Reeves won Game Number 200 during the 2003 NFL season; his team won one more gain before he was replaced as head coach of the Atlanta Falcons. How many more wins does Don Shula have than Dan Reeves?

_____

Challenge problem

Why does Halas have so many more tie games than Shula?

_____

From *Football Math: Touchdown Activities and Projects for Grades 4–8* published by Good Year Books. Copyright © 1995, 2005 Jack Coffland and David A. Coffland.

Home Cooking—Different Flavors

Most teams expect to win more games at home and fewer on the road. But that is not always true. Examine the home and away records of the Eagles, Vikings, Colts, and Chiefs during a four-year time period.

Home and Away Records for 2000, 2001, 2002, and 2003

Team	Home Wins	Home Losses	Home Winning %	Road Wins	Road Losses	Road Winning %
Eagles	21	11		25	7	
Vikings	22	10		9	23	
Colts	19	13		19	13	
Chiefs	22	10		12	20	

1. Figure the winning percentage for both home and away games for each team and enter them into the chart. (Round percent answers to whole numbers.)

2. Then, for each team, compare the home winning percentage with the away winning percentage and complete the following chart.

Team	Home Winning %	Road Winning %	Percentage Difference
Eagles			
Vikings			
Colts			
Chiefs			

3. Which team had the biggest positive difference between the home and road winning percentage? (Or, they won a lot more games at home than away.)

4. Which team had no difference between the home and road winning percentages? (Or, their home and away records were the same.)

5. Which teams had the best home winning percentage?

6. Which team had the worst road winning percentage?

Comparing the "Best of the Best" Coaches

At the end of the 2003 season, only five of the seven coaches with 200+ wins had 200 or more wins during the regular season. Let's examine their records in more detail.

NFC Coaches with 200+ Wins During the Regular Season

Coach	Years as Head Coach	Games Won	Games Lost	Games Tied
Don Shula	33	328	156	6
George Halas	40	318	148	31
Tom Landry	29	250	162	6
Curly Lambeau	33	226	132	22
Paul Brown	25	213	104	9

1. Compute the average number of "wins per season" for each coach and put that figure in the chart below.

2. Then compute the percentage of games won for each coach. (*Note:* Figure the percentage of games won out of all the games coached by these individuals. You have to add the wins, losses, and ties to find the total number of games each man coached in the NFL.) Then put those figures in the chart below.

Coach	Average Wins Per Season	Percentage of Games Won
Don Shula		
George Halas		
Tom Landry		
Curly Lambeau		
Paul Brown		

Challenge problem

Is it fair to compare the number of victories in a season for each of the coaches, considering their records span more than 80 years of NFL history? Why or why not?

_____

_____

Or, do you feel it is better to compare the percentage of games won to games played in order to compare these coaches?

_____

From *Football Math: Touchdown Activities and Projects for Grades 4–8* published by Good Year Books. Copyright © 1995, 2005 Jack Coffland and David A. Coffland.

Monsters of the Midway

Solve the following problems.

1. Walter Payton holds the all-time season rushing mark for the Chicago Bears. He rushed for 1,852 yards in 1977. In those days teams only played 14 games in a season. What was the average number of yards that Payton gained each game?

_____

2. Walter Payton holds the Chicago Bear record for yards gained in one game. On November 20, 1977, he rushed for 275 yards in a single game. If he were to gain that many yards during every game using today's schedule (teams play 16 games now), how many total yards would he gain in a season? (*Note:* This imaginary figure would double the existing season rushing record for a running back!)

_____

Quarterback	Date	Yards Gained
Johnny Lujack	12/11/1949	468
Bill Wade	11/18/1962	466
Sid Luckman	11/14/1943	433

3. The Chicago Bears are one of the few NFL teams that did not set new passing records during the 1980s when the rules were modified to make defending against passing more difficult. The chart shows that the records for the top three passing games by a Chicago QB are all more than 40 years old. How many total passing yards did the Bears have in their top three passing games of all time?

_____

Challenge problem

Can you find information about these famous Chicago quarterbacks? When did they join the Bears? When did they leave? Are any of the players in the NFL Hall of Fame?

_____

_____

_____

_____

4. Kevin Butler, who kicked field goals for Chicago from 1985–95, holds the career record for most field goals for the Bears. He kicked 243 field goals during his career. How many points did he score on field goals?

_____

5. Walter Payton holds many Chicago Bear rushing records, but he does not hold the records for touchdowns scored by a running back, either in a season or a game. Gale Sayers, a famous running back from the 1960s, scored 6 touchdowns in one game (he is still tied for the NFL record in that category) and 22 touchdowns in a season. A touchdown is worth 6 points; how many points did Sayers score in his record game? How many points did he score in his team-leading season?

Game: _____

Season: _____

From *Football Math: Touchdown Activities and Projects for Grades 4–8* published by Good Year Books. Copyright © 1995, 2005 Jack Coffland and David A. Coffland.

St. Louis Rams Records

The St. Louis Rams have a long, interesting history in the NFL. The franchise has been located in three cities: Cleveland, Los Angeles, and St. Louis. They have had many famous players, including Norm Van Brocklin, Elroy (Crazylegs) Hirsch, Tank Younger, Merlin Olson, and pass rusher Deacon Jones. The table below shows some team records of the Rams.

St. Louis Rams Team Records

Record	Player, Year(s)	Yards	Record?
Career Rushing Yards	Eric Dickerson, 1983–87	7,245	
Career Passing Yards	Jim Everett, 1986–1993	23,758	
Career Receiving Yards	Henry Ellard, 1983–1993	9,761	
Season Rushing Yards	Eric Dickerson, 1984	2,105	NFL Record
Season Passing Yards	Kurt Warner, 2001	4,830	
Season Receiving Yards	Issac Bruce, 1995	1,781	

1. Eric Dickerson played for the Rams from 1983 through 1987, a total of five seasons. How many yards did he gain on the average each season?

2. Issac Bruce set the modern Ram records for receiving yards over an entire season. His record of 1,781 yards was made in the current 16-game season. How many yards did he average per game during that 1995 season?

3. Elroy "Crazylegs" Hirsch held the old Ram record for yards gained on pass receptions in one season. His record was set during a 12-game season in 1951, when he gained 1,425 yards in pass receptions. What was his average "yards gained per game" for the 1951 season? Because Crazylegs' average per game is higher than that of Issac Bruce, what would the Rams' record be if Crazylegs had played 16 games with his "average yards gained per game" amount during that 1951 season?

From *Football Math: Touchdown Activities and Projects for Grades 4–8* published by Good Year Books. Copyright © 1995, 2005 Jack Coffland and David A. Coffland.

St Louis Rams Records (cont'd.)

Older Ram Records			
Record	**Player, Year(s)**	**Yards**	**Record?**
Game Passing Yards	Norm Van Brocklin 9/28/1951	554	NFL Record
Game Receiving Yards	Willie Anderson 11/26/1989	336	NFL Record
Game Receptions	Tom Fears, 1950	18 catches	NFL Record

4. Norm Van Brocklin's NFL and Ram record for yards passing in one game has been on the record books since 1951. Using this year's date, how long has his record lasted?

5. Willie Anderson's individual game record for yards gained receiving is an exceptional number. If he could average that over all 16 games in an NFL season, how many yards would he gain on pass receptions?

Challenge problem

Tom Fears caught 18 passes in one game during the 1950 NFL season. That is the NFL record for pass receptions in one game. If he had caught that many passes during all 12 games his team played that year, how many passes would he have caught during that season? Would that figure be an NFL record even now, when teams play 16 games in a season? Could you have predicted whether your number (projecting a single game record to be a player's average for every game of the year) would be an NFL record?

From *Football Math: Touchdown Activities and Projects for Grades 4–8* published by Good Year Books. Copyright © 1995, 2005 Jack Coffland and David A. Coffland.

Cleveland Browns Heroes

The Cleveland Browns were originally part of the old All American Football Conference in the 1940s. The Browns were admitted to the NFL in 1950 when the NFL and All American Conference merged. Cleveland came into the NFL with a great team and several Hall of Fame players.

1. Cleveland won six conference titles and three league titles during the team's first six years in the league. During those six years, they won 58 games. What was their average number of wins per year during those first six seasons, rounded to two decimal places?

2. Otto Graham was the quarterback for those Cleveland teams. He averaged 8.63 yards for each pass attempt in his career; this is still the NFL record. He attempted 1,565 passes. How many total yards did he gain in all of those attempts?

3. Jim Brown was a running back for Cleveland from 1957 through 1965. When he retired, he held virtually every NFL career rushing record. He gained 12,312 yards in the nine years he played. Walter Payton of the Chicago Bears finally broke Brown's record in the 1980s. Payton gained a total of 16,726 yards rushing. How many more yards did Payton gain than Jim Brown? (*Note:* Payton played 13 years, or 4 years longer than Brown.)

4. Jim Brown gained his 12,312 yards in 9 seasons. Payton gained his 16,726 yards during 13 seasons. Who gained more yards on the average per season? (*Note:* Jim Brown played fewer games per season.)

Challenge problem

Probably the most famous field goal kicker in the NFL's early years, Lou Groza, also played for those early Cleveland Browns. Groza still holds two kicking records for Cleveland—the total points scored and the total number of field goals made. Groza kicked 234 field goals and scored a total of 1,349 points. Can you use these two numbers to find out how many "extra point kicks" he made for the Browns?

Cincinnati Bengals History

The Cincinnati Bengals were formed in 1968 as an expansion team in the AFL. Paul Brown, the famous Cleveland Browns coach, was part owner and coach of the Bengals. He coached the team for eight years, compiling a record of 55 wins, 59 losses, and 1 tie. In their third season Brown coached the team to a record of 8 wins and 6 losses, and they made the play-offs for the first time. They lost to the Baltimore Colts (the eventual Super Bowl Champion that year) by a score of 17–0. Sam Wyche, who coached the Bengals from 1984 to 1991, led the team to more wins than any other Bengal coach. His record with the Bengals was 64 wins and 68 losses. He took Cincinnati to Super Bowl XXIII, but they lost to San Francisco by a score of 20 to 16. Recently, Bengal fans have suffered through many losing seasons. In fact, from 1998 to 2002, the Bengals lost 10 games or more every season. The team won only 19 games while losing a total of 61 games in those 5 seasons. In 2003 they hired Melvin Lewis as head coach, and the team immediately won half their games. It was the first time since the 1996 season that the Bengals did not finish the season with a losing record.

1. How many points were scored in Super Bowl XXIII?

   _____

2. What was the total number of games that Paul Brown coached for the Bengals?

   _____

3. If Paul Brown won a total of 170 games, how many games did he win with other teams?

   _____

4. How many more games did Sam Wyche coach for the Bengals than Paul Brown?

   _____

5. The Bengals' record from 1998 to 2002 was not good. How many more losses than wins did they have over those five seasons?

   _____

6. As of the 1992 season, Cincinnati had won 184 games in the NFL. How many games were won by coaches other than Paul Brown and Sam Wyche?

   _____

From *Football Math: Touchdown Activities and Projects for Grades 4–8* published by Good Year Books. Copyright © 1995, 2005 Jack Coffland and David A. Coffland.

Indianapolis Colts History

The Indianapolis Colts did not always play in Indianapolis. In fact, they had a much longer history as the Baltimore Colts. Therefore, many of their team records are held by players who played in Baltimore.

1. The most famous Colt of all was Johnny Unitas, the quarterback who played for Baltimore from 1956–72. During his 17 years in the NFL, "Johnny U" passed for 39,768 yards. How many yards gained passing did he average during each of those 17 years?

_____

2. Peyton Manning joined the Colts in 1998; like Johnny Unitas, he is on his way to becoming one of the "best" quarterbacks in NFL history. He now holds the Colts single-season passing record. During the 2000 season, he completed 357 passes for a total of 4,413 yards. What was his average "yards per completion"?

_____

3. Edgerin James joined the Colts in 1999 as a running back; he quickly set a new Colt record for yards gained rushing in a single season. In 2000 James rushed for a total of 1,709 yards; he carried the ball 387 times to set that record. What his average "yards gained per carry" figure?

_____

4. Marvin Harrison joined the Colts in 1996 as a wide receiver. He broke the team's long-standing record for passes caught and yard gained receiving. Both team records were set in 2002; the 143 passes Harrison caught that season is also the NFL record for receptions made in a single season. Harrison's receiving record is 1,722 yards gained in 16 games. What was his "average receiving yards per game" figure?

_____

5. Raymond Berry was a wide receiver for the old Baltimore Colts; he was the favorite receiver of Johnny Unitas. He held the Colts' season record for pass receiving yardage until Harrison broke his record in 2002. During the 1960 season, Berry gained 1,298 yards catching passes in a 12-game season. How many yards did he average for each of the 12 games?

_____

Lincoln's Big Win

Solve the following problems.

1. The Lincoln Junior High football team scored a big win over Central Junior High. The quarterback passed for 325 yards, and the running backs ran for 279 yards. How many total yards did Lincoln gain on its way to its easy victory?

_____

2. Lincoln's star wide receiver caught eight passes for a total of 256 yards. What was his average gain for each of the eight receptions?

_____

3. Central Junior High did not have a very good day. They only gained 136 yards total, and 48 of these came on one good pass play. How many yards did Central gain on all the rest of their plays?

_____

4. After a winning game, the Lincoln cheerleaders do a pushup for each point the team scored. Lincoln has 12 cheerleaders. If they each did 35 pushups, how many total pushups did they have to do at the end of the game?

_____

5. Lincoln let every person on its team play in an effort to keep from running up the score. If they have 37 offensive players and 39 defensive players, how many players did Lincoln play in the game?

_____

6. The game was played on a very hot day. Each of Lincoln's players drank five glasses of water during the game. How many cups of water did the trainers have to fill during the game? (_Hint:_ Where can you find the total number of football players on Lincoln's team?)

_____

From Football Math: Touchdown Activities and Projects for Grades 4–8 published by Good Year Books. Copyright © 1995, 2005 Jack Coffland and David A. Coffland.

Halftime Munchies

Examine the following menu. Then find what your bill would be for each meal purchased at the stadium. Remember that you may have to make several calculations to find the answer.

Menu

Hot dogs	$2.00
Polish sausage dog	$3.00
Nachos	$2.50
Pizza (slice)	$2.00
Small soft drink	$1.00
Medium soft drink	$1.50
Large soft drink	$2.00
Souvenir cup	$3.00
Large coffee	$1.00

1. Mary Jo bought two hot dogs, a small soft drink, and two medium drinks. How much was her total purchase?

2. David bought a Polish sausage dog, two orders of nachos, and a large soft drink. He ate it all so fast he felt terrible for the rest of the game. How much did he have to pay for the meal that made him feel so bad?

3. When Don went to get something to eat, the score was Kansas City 24, Denver 14. There was a long line at the concession stand. When he got back, Denver was ahead by 3 points. Kansas City did not score while he was gone. Find the total number of points on the scoreboard when Don came back to his seat.

4. Karen went to buy food for everybody in her group. She insisted that they all buy the same thing because she couldn't remember seven different orders. So she bought seven hot dogs and seven small colas at the Concession Stand. How much did she spend in all?

5. Trisha had a 10-dollar bill for food. If she bought a Polish sausage dog, one order of nachos, and a souvenir drink, how much change did she receive?

From *Football Math: Touchdown Activities and Projects for Grades 4–8* published by Good Year Books. Copyright © 1995, 2005 Jack Coffland and David A. Coffland.

All-purpose Yards Leaders I

An interesting new statistic that is being kept for football players is "all-purpose yards gained." This statistic is kept for players who may carry the ball, catch passes, return punts, or return kickoffs. The NCAA record for most all-purpose yards in one game was set in 2000 by Emmett White of Utah State. He had 322 rushing yards, 134 receiving yards, 2 yards on punt returns, and 120 kick return yards, for a total of 578 all purpose yards.

Use the information presented in the following problems to compute the "all-purpose yards gained" statistic for each of the players described in the problem. Remember that you may have to make several computations to find the total yards gained.

1. "Flash Frederickson" averaged 9 yards per carry for 14 running plays. He also returned a kickoff for 38 yards and caught two passes for a total of 67 yards. How many all-purpose yards did Flash gain in this ball game?

2. "Stonehands" Harrison is a wonderful football player, but they call him "Stonehands" because he fumbles a lot. Last week "Stonehands" had an average game. He gained 123 yards running, caught two passes for 42 yards, and returned three kickoffs for exactly 24 yards each time. How many all-purpose yards did "Stonehands" gain that day? (Unfortunately, it was an average game; he also had four fumbles!)

3. Jimmy "The Flea" Flicker is a little player who is very fast. He caught five passes for an average of 18 yards, returned six punts for an average of 23 yards, and returned two kickoffs for an average of 42 yards. How many all-purpose yards did he gain?

Challenge problem

Can you find statistics on how many all-purpose yards a player gained in a college or professional football game that occurred this year?

From *Football Math: Touchdown Activities and Projects for Grades 4–8* published by Good Year Books. Copyright © 1995, 2005 Jack Coffland and David A. Coffland.

College Quarterback Ratings

The quarterback position is so important for a football team. How do you tell who is the best quarterback? People who create football statistics have tried different ways of rating passers to give some comparative indication of their ability. But many things can happen when a quarterback throws the ball, so the statistical formulas are rather complicated.

The formula for rating college quarterbacks is shown below. An example is provided to help you work your way through the statistic; a second problem is given for you to do on your own.

Wild Willie Wilson's Passing Statistics for Last Year

Passes completed:	143
Passes attempted:	251
Yards gained:	1,543
Passing touchdowns:	21
Interceptions:	10

Step 1: The pass completions are divided by the attempts; the answer is multiplied by 100.

$$143 \div 251 = .5697 \times 100 = 56.97$$

Step 2: The yards gained passing are divided by the attempts; the answer is multiplied by 8.4.

$$1543 \div 251 = 10.79 \times 8.4 = 90.636$$

Step 3: The number of touchdowns is divided by the attempts; the answer is multiplied first by 100 and then by 3.3.

$$21 \div 251 = .08366 \times 100 = 8.366 \times 3.3$$
$$= 27.609$$

Step 4: The interceptions are divided by the attempts; the answer is multiplied by 100 first and then by 2.

$$10 \div 251 = .0398 \times 100 = 3.984 \times 2$$
$$= 7.968$$

College Quarterback Ratings (cont'd.)

Finally, add the answers for Steps 1, 2, and 3, then subtract the answer for Step 4.

$$56.97 + 90.636 + 27.609 =$$
$$175.215 - 7.968 = 167.247$$

So, Wild Willie Wilson's rating is 167.247. This statistic is calculated to make 100 the average score for quarterbacks. So, Wild Willie is an above average quarterback. In fact, he looks to be a very good quarterback.

Now, try this statistic one more time. Below you will find the astatistics for Sad Sam Smith. (*Hint:* You might guess from Sam's nickname that he is a below-average quarterback!)

Sad Sam Smith's Passing Statistics

Passes completed	98
Passes attempted	211
Yards gained	843
Passing touchdowns	8
Interceptions	19

From *Football Math: Touchdown Activities and Projects for Grades 4–8* published by Good Year Books. Copyright © 1995, 2005 Jack Coffland and David A. Coffland.

Forward and Backward

Figuring an offensive team's total yardage gained is not just a case of adding up the numbers. Players make positive plays, such as yards gained rushing and yards gained passing. But they also make negative plays, such as yards lost rushing or yards lost on a quarterback sack. There are also penalties; they work both ways. Examine the following game situations and figure the total yards gained by the team.

1. The East Side Buffaloes gained 256 rushing yards in a game. They also lost 4 yards on one play. The quarterback threw for 154 yards, but he was sacked seven times. He lost a total of 53 yards on the sacks. How many total yards did the Buffaloes gain?

_____

2. In one game the Southside Stinkers gained 187 yards rushing and 458 passing. But they had 211 yards of penalties against the offense. How many total yards did they gain that night?

_____

3. The Boise Bullets are a passing team. They ran for only 47 yards last night, but they passed for 398. They lost 38 yards on sacks, and they had three offensive penalties against them for a total of 25 yards. How many total yards did they gain last night?

_____

4. The Lament Lame Ducks didn't win a game from 1991 through 1993, but last night they did win. The Lame Ducks only gained 75 yards passing and 62 yards rushing, but their opponent had 165 yards in defensive penalties. When you count all of their yards and all of the penalty yards, how many yards did the Lame Ducks gain? (Even after all of those gift yards, the Lame Ducks only won by two points!)

_____

Forward and Backward (cont'd.)

5. Marta Lane, the principal at Horseshoe Bend High School, was trying to explain why they won a game so easily. She knew her team had run for 387 yards and passed for 256 yards. She also knew that they had been sacked only one time for a loss of seven yards, and only one penalty had been called against the offense for just five yards. But she didn't know how to figure the total yardage gained, so she simply said, "We marched up and down the field all night." Can you help Marta figure the total yards gained by her team that night?

_____

6. Bob's favorite team was run out of the park last Saturday. They gained only seven yards rushing and 46 yards passing. They were penalized six times for 40 yards and they lost 72 yards when the quarterback was sacked nine times. What was the total yardage figure for Bob's team? (Is it possible to find an answer this problem? How would you explain it?)

_____

_____

From _Football Math: Touchdown Activities and Projects for Grades 4–8_ published by Good Year Books. Copyright © 1995, 2005 Jack Coffland and David A. Coffland.

Football Fashions

A large number of people watch football games. They pay a lot of money to attend each game. Teams need to attract large crowds, because it costs a great deal of money to put a football team on the field. For example, the Louisville Tigercats, which has 35 team members, paid the following equipment prices in 1955. (Equipment costs more now!)

Football Equipment Prices, 1955 Catalog

Equipment	Unit Price	Needed	Total
Helmet	$35	35	
Shoulder Pads	$28	35	
Hip Pads	$18	35	
Thigh Pads	$ 4	70	
Knee Pads	$ 3	70	
Shoes	$45	35	
Jersey	$18	90	
T-shirt	$ 8	90	
Pants	$26	90	
Belt	$ 5	35	
Socks (pairs)	$ 2	92	

1. What is the total cost for each item, given the quantity of each item that the Tigercats needed?

_____

2. The Tigercats bought more than one jersey for each player, because jerseys are torn during the games. If one-third of the Tigercat jerseys were torn during the season, how much did it cost to replace them?

_____

Football Fashions (cont'd.)

3. One-seventh of the Tigercats' shoulder pads were broken due to hard hitting. How much will it cost to replace them?

_____

4. The Tigercats wore out three-quarters of their socks. How much will it cost to replace them?

_____

5. One-fifth of the Tigercat helmets were damaged so badly that they had to be thrown away. How many helmets did the Tigercats have left after the season?

_____

6. One-fifth of the Tigercats' uniform pants and one-seventh of their hip pads needed to be replaced after the season was over. What was the cost for these replacements?

_____

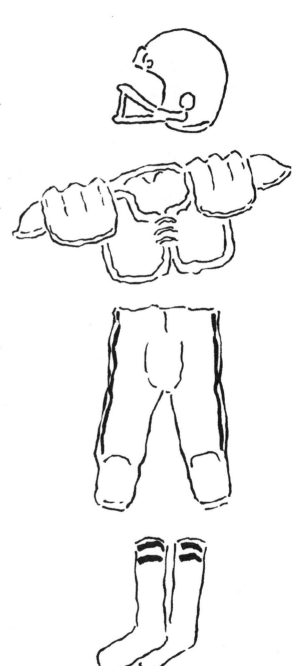

New York Jets Records

The New York Jets began as part of the American Football League; they entered the NFL when the two leagues merged. They played in the Super Bowl III, that famous game in which Joe Namath literally "guaranteed" a victory over the Baltimore Colts. The Jets won the game, making it the first time an American Conference team beat a National Conference Team.

1. Pat Leahy holds most of the New York Jet career kicking records. He kicked a total of 304 field goals, a Jet record. He also scored 1,470 points over his career, also a Jet career record. Using these figures, can you find how many extra points Leahy kicked?

_____

2. Joe Namath held the Jets' record for TD passes in a single season for many years. He threw for 26 touchdowns in 1967. Vinny Testaverde holds the record today; he threw for 29 TDs in 1998. How many more points were scored on Testaverde's passes? (Remember not to count extra points!)

_____

3. Don Maynard, who played for the Jets from 1960–72, holds the Jets' record for total career touchdowns scored; he is also the co-holder of the record for season touchdowns scored. If his record for a career is 88 touchdowns, and he scored 14 of them during the year 1965 in which he set the season record, how many points did he score on his touchdowns for all of the other years he played?

_____

4. Actually, three different players are tied for the Jets' record for TDs scored in a single season. Art Powell (1960), Don Maynard (1965), and Emerson Boozer (1972) all scored 14 TDs in the seasons shown. How many points did these players score together?

_____

5. Curtis Martin has been the Jets' leading running back for several years. He entered the NFL in 1995, and he has gained more than 1,000 yards in each of his first nine seasons (1995 season through the 2003 season). In those first nine seasons, he gained 11,669 yards. How many rushing yards gained did he average over those nine seasons?

_____

Challenge problem

Did Martin continue this string of 1,000+ yards after the 2003 season? If so, for how many additional seasons? What is the record for consecutive seasons rushing for more than 1,000 yards?

_____

Buffalo Bills Records

Solve the following problems.

1. Jim Kelly holds the career passing record for Buffalo; he threw for 25,467 yards. He is in the NFL Hall of Fame. Thurmond Thomas holds Buffalo's career rushing record; he ran for 11,938 yards. He certainly has the potential to be in the Hall of Fame. How many yards did these famous players gain altogether?

_____

2. Paul Macguire was a punter for the Buffalo Bills. He holds Buffalo's single-season record for the highest average yards per punt. He set the record in 1969, averaging 44.5 yards per punt. If he punted 37 times, what was the total yardage for all of his punts?

_____

3. Eric Moulds was Buffalo's best receiver between 1995 and 2003. He holds Buffalo's record for most receptions in a season (94, set in the year 2000) and most yards gained receiving in a season (1,368, set in 1998). If he had set both records in the same year, what would his "average yards per catch" statistic be?

_____

4. Butch Byrd, who played defensive back for the Bills from 1964–70, holds the all-time Bills record for interceptions. He intercepted 40 passes during the seven years that he played for the Bills. Assume that the other teams would have scored 12 field goals and 9 touchdowns on their drives if Byrd had not intercepted the passes. (The other times they would have punted the ball.) How many points did he prevent with his interceptions?)

_____

5. Drew Bledsoe was traded from New England to Buffalo after the 2001 NFL season. He played for nine seasons in New England, passing for a total of 29,657 yards. In his first two seasons with Buffalo, Bledsoe threw for an additional 7,219 yards. What is his total number of passing yards gained for his career with both teams?

_____

Challenge problem

Dan Marino of the Miami Dolphins holds the career "total yards" passing mark for NFL quarterbacks. Can you find his record? How far away from the record is Drew Bledsoe? (Remember: You found Bledsoe's numbers in problem 5.)

_____

From *Football Math: Touchdown Activities and Projects for Grades 4–8* published by Good Year Books. Copyright © 1995, 2005 Jack Coffland and David A. Coffland.

Dallas Cowboy's Super Bowls

Bowl	Winner	Score	Opponent	Score	Winner	Loser
SB V	Baltimore	16	Dallas	13	$15,000	$7,500
SB VI	Dallas	24	Miami	3	$15,000	$7,500
SB X	Pittsburgh	21	Dallas	17	$15,000	$7,500
SB XII	Dallas	27	Denver	10	$18,000	$9,000
SB XIII	Pittsburgh	35	Dallas	31	$18,000	$9,000
SB XXVII	Dallas	52	Buffalo	17	$36,000	$18,000
SB XXVIII	Dallas	30	Buffalo	17	$36,000	$18,000
SB XXX	Dallas	27	Pittsburgh	13	$42,000	$27,000

1. The Dallas Cowboys have played in the Super Bowl eight times, more than any other team. They played their first Super Bowl game in 1971, when they played Baltimore and lost. They played their most recent Super Bowl game in 1996. That game was against Pittsburgh, and they won. How many years were there between Dallas's first Super Bowl game and their most recent?

   _____

2. How many total points has Dallas scored in its Super Bowl games?

   _____

3. How many points have Dallas opponents scored in those same games?

   _____

4. What was the average number of points scored by Dallas? Their opponents?

   _____

Dallas Cowboy's Super Bowls (cont'd.)

5. Assume that Dallas had one coach for all of their Super Bowl games. How much would that coach earn for coaching in those Super Bowls? (Remember that he would get the winner's share if Dallas won the game, and he would get the loser's share if they lost the game.)

_____

6. If 53 Dallas players were given the winner's share in 1996, then what was the total amount of money paid to the Dallas players?

_____

From _Football Math: Touchdown Activities and Projects for Grades 4–8_ published by Good Year Books. Copyright © 1995, 2005 Jack Coffland and David A. Coffland.

From Football Math: Touchdown Activities and Projects for Grades 4–8 published by Good Year Books. Copyright © 1995, 2005 Jack Coffland and David A. Coffland.

Detroit Lions Stars

Barry Sanders, the great Detroit running back of the 1990s, is tied for second on the list of players who have led the league in rushing. Jim Brown won the rushing title during eight different seasons; Sanders and four other players are in second place, having led the league in rushing four times. The chart below shows Sanders's statistics for leading the league in rushing:

Barry Sanders
Years Leading the NFL in Rushing

Year	Total Yards Gained Rushing
1990	1,304
1994	1,883
1996	1,553
1997	2,053

1. How many total yards did Sanders gain in his years of leading the league in rushing?

_____

2. Sanders gained a total of 15,269 yards rushing over his career. How many total yards did he gain during the years he did not lead the league?

_____

Jason Hanson has kicked field goals for Detroit since 1992. He is high on the list of successful field goal kickers, both for his accuracy and for his length. He is high on the list of two field goal records.

3. Hanson is tied for fourth on the list of kickers making field goals of 50 yards or longer; he has accomplished the task 21 times. How many points were scored on his 50 yards or longer field goals?

_____

4. Hanson is also one of 19 kickers tied for second place for field goals made in a game. Three kickers have made 7 field goals in a game; 19 have made 6 in a game. What was the total number of points scored by the 22 kickers when all of their points are combined?

_____

Kansas City Chiefs History

The Kansas City Chiefs played their first game during the 1960 season as part of the old American Football League. Then known as the Dallas Texans, they played in Dallas for the 1960, 1961, and 1962 seasons before moving to Kansas City.

1. Kansas City has been involved in the two longest games in league history. The longest game was a playoff game against Miami on Christmas Day 1971. At the end of the game, the score was ties at 24 points. Neither team scored in the first overtime period. Miami finally made a field goal 7 minutes and 40 seconds into the second overtime period. If each quarter and the first overtime were 15 minutes long, what was the total length of that football game?

_____

MISSOURI

2. In the second longest overtime game, the Dallas Texans were playing the Houston Oilers in the 1962 AFL championships game. The wind was blowing very hard. Dallas won the coin toss and wanted to take the wind at their back. The captain made a mistake, however, and Houston got both the ball and the wind. But Houston did not score in the first overtime. Dallas then got the wind in the second overtime, and they kicked a field goal 2 minutes and 54 seconds into the second overtime. What was the total length of that game? (Again, each quarter and the first overtime period were 15 minutes long.)

_____

3. Kansas City also played in the first Super Bowl, where they met the Green Bay Packers on January 15, 1967. Green Bay won the game, 35 to 10. If only one field goal was kicked during the game, what was the total number of TDs scored by both teams?

_____

4. Three years later, in 1970, Kansas City returned to the Super Bowl and defeated the Minnesota Vikings 23 to 7. Minnesota threw three interceptions and lost two fumbles. If all of those Minnesota possessions might have resulted in touchdowns, what was the maximum number of points that Kansas City prevented with the turnovers?

_____

From *Football Math: Touchdown Activities and Projects for Grades 4–8* published by Good Year Books. Copyright © 1995, 2005 Jack Coffland and David A. Coffland.

Crazy College Capers

Solve the following problems

1. On November 12, 1955, a raging storm blew across the rolling hills of eastern Washington. Unfortunately, Washington State University had scheduled a game with San Jose State that afternoon. It was cold and snowy, and a fierce wind blew. The game went on anyway. Although 412 students came to the game, they attend games for free. For a long time, no adult came to buy a ticket. Finally, one lonely fan bought a ticket to see the day's contest. After that, the ticket sellers should have gone home. No one else bought a ticket that day—a good thing, because it snowed over a foot during the game, making visibility poor. It was difficult to tell if a player had made a first down or not. Did it matter? No. The game ended in a 13–13 tie. How many more points were scored than tickets sold?

_____

From *Football Math: Touchdown Activities and Projects for Grades 4–8* published by Good Year Books. Copyright © 1995, 2005 Jack Coffland and David A. Coffland.

Crazy College Capers (cont'd.)

2. During the Cotton Bowl in 1954, the Rice Owls were playing the Alabama Crimson Tide. The Owls had the ball, deep in their own territory at their own 5-yard line. The quarterback pitched the ball to Dickie Moegle, the star Rice halfback. Moegle broke free around the right end. He was off to the end zone, 95 yards away. He was running free down the Alabama sideline. He was past the last Alabama player; he would score! Suddenly Tommy Lewis, an Alabama captain, was there, without a helmet. He made a beautiful open field tackle at the Alabama 41-yardline. There was just one problem. Lewis was not on the field for this play. He had run from the bench to make the touchdown-saving tackle! The referees huddled. What should they do? Lewis obviously should be called for a penalty. Their decision: Moegle was running free; the rules state that he should be awarded a touchdown. A player cannot come off the sideline to illegally participate in a play. How far did Moegle actually run with the ball?

3. The most famous mix-up in football occurred in 1982 during the final seconds of the Stanford-California season-ending game. Stanford quarterback John Elway had just directed a scoring drive; Stanford was ahead 20 to 19 with only four seconds left to play. All Stanford had to do was kick off and tackle the runner, and the game would be over—a tremendous "comeback" win engineered by John Elway. But Stanford kicked short to prevent a runback. Kevin Moen caught the ball, starting forward. He was surrounded. He lateraled the ball to a teammate. That player was surrounded. He lateraled to another teammate just before being tackled. The clock reached zero. The Stanford fans cheered. The Stanford band started marching out of the end zone onto the field. But the ball was still live. A final lateral put the ball back in Kevin Moen's hands. He ran into the Stanford band. Was he tackled? Almost. By whom? A Stanford trombone player in the end zone! California had scored the winning TD, with the help of Stanford's band. California's best blockers were the Stanford band members. What was the final score?

Solve the following problems.

1. The Pep Club is decorating for the High School Homecoming Game. The club got a bargain price of $3 a yard for cloth to make banners. If they bought all 35-1/3 yards of material that the store had, how much did they have to pay?

_____

2. Club members cut the 35-1/3 yards of cloth into banners that were two-thirds of a yard long. How many banners did they make?

_____

3. Jeff and Juanell were in charge of making signs. They each got a piece of paper from the office; Juanell's was 4-2/3 yards long and Jeff's was 3-1/2 yards long. If they combined their paper to make a longer sign, how long would it be all together?

_____

4. Sam cut 5-1/3 yards of crepe paper into strips that were 3/4 of a yard long. After he was finished, he decided that he needed more strips, so he got another 5 strips from LaShonda. How many strips did he have when he was done?

_____

5. Karen had 15-1/2 yards of streamers. She used half of it to make one decoration. Then she used 2-1/3 yards for another decoration. How much of her original piece of material is left?

_____

Quarterback Stats

League statistics for the Frontier League were released yesterday. They indicate the percentage completion figures for all of the quarterbacks in the league. Only part of the statistics are shown below; work with the numbers to complete the figures. Remember that all percentage completion figures are rounded off, so you will have to round off all of your answers to the nearest whole number. (After all, you can't complete half of a pass.)

Frontier League Quarterback Statistics

Quarterback & Team	Passes Attempted	Percentage Completed	Passes Completed
Tom Davis, Bengals	148	52%	
Mike Irving, Broncos	91		38
Jimmy Anderson, Vandals		47%	58
Don Buratto, Bantams	156		72
Steve Johnson, Grayhounds		41%	43
Bob Myers, Bulldogs	112	45%	

1. The table shows that Tom Davis threw 148 passes. He completed 52% of them. How many passes to his receivers did he complete?

   _____

2. What percentage of his passes did Mike Irving complete?

   _____

3. How many passes did Jimmy Anderson attempt?

   _____

4. Dan Buratto completed what percentage of his passes?

   _____

5. How many passes did Steve Johnson attempt?

   _____

6. How many passes did Bob Myers complete?

   _____

7. From these statistics, who do you think should be the all-star quarterback? Why?

   _____

Frantic Fans

Solve the following problems.

1. Jerry and Barbara collect football cards. Jerry tried to give Barbara one card so that they will both have the same number of cards. But Barbara said, "No, I will give you your card back and one of my cards." Now Jerry has twice as many cards as Barbara. How many cards did each begin with?

2. The sixth-grade San Diego Chargers' Fan Club at Coral Ridge Elementary School collected dues from all of its members. The club collected $2.89. If each club member paid the same amount, and each club member paid that amount with five coins, how many nickels did the club receive?

3. Kyle and Conner have three favorite players on their home team. They like to know every statistic about their favorites. Kyle saw that when he multiplied his three favorite players' years of experience together, the answer was 24. Conner saw that when he added the players' years of experience, he got an answer of 11. How much experience did each player have?

From *Football Math: Touchdown Activities and Projects for Grades 4–8* published by Good Year Books. Copyright © 1995, 2005 Jack Coffland and David A. Coffland.

How Big Is the Team?

Solve the following problems.

1. Prescott is a small high school. The Prescott Tigers have 15 players for offense and 16 players for defense. However, 5 players play both offense and defense. How many players are on the Prescott Tigers football team?

_____

2. The Cougars are from a very large high school. Their offensive team has 34 players, their defensive team has 42 players, and their special teams have 24 players. However, 11 players play both offense and defense, 10 players play defense and special teams, and 6 players play offense and special teams. Three players are on all 3 squads. How many players are on the Cougars?

_____

Rapid Runners

Solve the following problems.

1. The Tigers play an eight-game season. After three games in the season, Cesar has gained 387 yards running with the football. Is he on pace to gain 1,000 or more yards rushing for the season?

_____

2. During his junior year, Jorge ran for 758 yards. In his senior year he rushed for 1,093 yards. What was his percentage improvement from his junior to his senior year?

_____

Winner's Edge

Solve the following problems.

1. Maria challenged Bob to find how many games the Denver Broncos won during the 1980s. They played in the Super Bowl several times during the 1980s, she said, so they had a good team. Then she gave Bob the following clues to find the number:

   a. It is an odd number.
   b. It is greater than 9×9.
   c. The sum of the numeral's digits is an even number.
   d. It is a multiple of 3.
   e. It is less than 99.

   Can you help Bob answer Mary's tricky question?

   _____

2. The final score for a college game was Cougars 19, Bruins 14. Football teams can score:

   2 points for a safety or for a 2-point conversion
   3 points for a field goal
   6 points for a touchdown with a missed extra point
   7 points with a touchdown and an extra point
   8 points for a touchdown and a 2-point conversion

List all of the possible combinations that could have made a final score of 19 to 14. Then pick two of the combinations that you think are most likely and explain why.

_____   _____

_____   _____

_____   _____

_____   _____

_____   _____

_____   _____

From *Football Math: Touchdown Activities and Projects for Grades 4–8* published by Good Year Books. Copyright © 1995, 2005 Jack Coffland and David A. Coffland.

Defensive Delights

Solve the following problems.

1. Defensive coaches keep track of the number of defensive tackles made by each player. They have developed a scoring system to help describe how each player did during a game. Defenders are awarded 2 points for a solo tackle and 1 point for an assisted tackle. If Dave the defensive tackle earns 16 points, and had twice as many assists as solo tackles, how many assists and how many solo tackles did he have?

2. Tank the Tackle can't remember his number. All he can remember is that the sum of the digits of his uniform number is a prime number and equals the first digit on his jersey. What is his number? (*Hint:* Defensive tackles can only wear numbers between 60 and 79.)

From *Football Math: Touchdown Activities and Projects for Grades 4–8* published by Good Year Books. Copyright © 1995, 2005 Jack Coffland and David A. Coffland.

Scoring Schemes I

Solve the following problems.

1. Karl the Kicker scored 135 points last year. He kicked both field goals and points after touchdowns (PATs). He scored four times as many points with field goals as he scored with PATs. How many field goals did he make; how many extra points did he make?

2. The Hokies scored seven times in their game with the Wahoos. They scored 33 points in all, making only TDs (7 points) and field goals (3 points). How many TDs and field goals did they score?

From *Football Math: Touchdown Activities and Projects for Grades 4–8* published by Good Year Books. Copyright © 1995, 2005 Jack Coffland and David A. Coffland.

Solve the following problems.

1. The Armadillos' quarterback, Toady Tom, is very fussy. He has to have exactly 4 cups of water halfway through a game. If Bill, the Armadillos' waterboy, only has containers that hold 5 cups and 3 cups respectively, how will he measure out 4 cups of water for Toady Tom to drink?

2. The Clemson Tiger mascot has a traditional job. After each Clemson score, the Tiger does the same number of push-ups as points his team has. (*Example:* If Clemson scores a TD and leads 7–0, the Tiger does 7 pushups. If Clemson then scores a field goal to make the score 10–0, the Tiger then does 10 more pushups, making a total of 17 pushups so far.) This pattern goes on until the game is over. (*Note:* The Tiger does not do 6 pushups for the TD and then 7 for the TD and the extra point.)

   In one game during the early 1980s, Clemson scored 84 points on 12 touchdowns. How many pushups did the Clemson Tiger do that day?

College and Professional Game Records

Both college and professional records have been kept for more than 75 years. The table below shows the college and professional game records for a number of categories.

College and Professional Game Records

Category (in 1 game)	Professional Record	College Record
Most Rushes	45	58
Most Rushing Yards	295	406
Most Rushing TDs	6	8
Most Receptions	20	23
Most Receiving Yards	336	405
Most Receiving TDs	5	6
Most Passes Attempted	70	83
Most Passes Completed	45	55
Most Passing Yards	554	716
Most Passing TDs	7	11
Most Points Scored	40	48
Most TDs Scored	6	8
Most FGs Scored	7	7

1. What percent of the college record is the professional record in the rushing yardage category?

_____

2. What percent of the college record is the professional record in the receiving yardage category?

_____

3. What percent of the college record is the professional record in the passing yardage category?

_____

4. Drew Bledsoe set the professional records for most pass attempts and most completions in the same game. What was his completion percentage for that game?

_____

5. Drew Brees set the college records for most pass attempts and most completions in the same game. What was his completion percentage for that game?

_____

Challenge problems

Can you find who set all of these records?

_____

Why are the college records higher than the professional records in all but one category?

_____

From *Football Math: Touchdown Activities and Projects for Grades 4–8* published by Good Year Books. Copyright © 1995, 2005 Jack Coffland and David A. Coffland.

1. In 1992 the Seattle Seahawks came close to setting an all-time record for the lowest number of points scored. Let's imagine they managed to score 10 points less than half as many points as the next lowest team, the New England Patriots. If New England had scored 186 points, how many would Seattle have scored?

_____

2. Sudden Sam the Scoring Machine wanted to make a touchdown in every game. In half of the games he played, he scored three TDs. In one-quarter of the games, he scored two TDs. In one-sixth of the games, he scored just one TD. In three games he did not score. Please answer the following questions:

a. How many games did Sudden Sam play in all together?

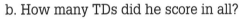

b. How many TDs did he score in all?

_____

c. What was the total number of points that he scored?

_____

Helpful Heroes II

Solve the following problems.

1. The Warriors are traveling to play the Wildcats. A windstorm was blowing as the Warriors left town. It took them 40 minutes at 30 miles per hour to get through the storm. Once the wind calmed down, the bus made up for lost time by averaging 70 miles per hour for 105 miles on the freeway.

   _____

   What was the average speed of the bus on the whole trip?

   _____

   If the team scheduled two hours for the trip, how many minutes were they early or late?

   _____

2. Carrie and Fred mixed caramel corn to sell in the concession stand. Because of a mix-up, they were given different recipes. Carrie's caramel corn was 70% popcorn and 30% peanuts, while Fred's was 60% popcorn and 40% peanuts. Together Carrie and Fred made a total of 30 pounds of caramel corn and used 19 pounds of popcorn. How much caramel corn did Carrie make? How much did Fred make?

   _____

From *Football Math: Touchdown Activities and Projects for Grades 4–8* published by Good Year Books. Copyright © 1995, 2005 Jack Coffland and David A. Coffland.

Projects

Average Salaries

Salaries for football players were lower than those for players in other sports until 1993. A new free-agency rule in that year allowed football players to increase their salaries dramatically. Below you will find the average salary for each football position.

Average Salaries for Each Position, 2003

Offense		Defense	
Quarterbacks	$2,143,008	Defensive Ends	$1,412,839
Running Backs	$992,157	Defensive Tackles	$1,304,462
Wide Receivers	$1,274,162	Linebackers	$1,343,450
Tight Ends	$884,049	Cornerbacks	$1,178,805
Offensive Linemen	$1,321,410	Safeties	$1,085,133
Kickers	$888,198	Punters	$784,680

Figure out the following.
(*Suggestion:* Because these are big numbers, you might like to use a calculator to help solve these problems.)

1. What is the average salary needed to field an offensive football team?

_____

2. What is the difference between the highest average salary and the lowest average salary?

_____

3. What is the total cost for an average defensive line? Offensive line?

_____

From *Football Math: Touchdown Activities and Projects for Grades 4–8* published by Good Year Books. Copyright © 1995, 2005 Jack Coffland and David A. Coffland.

From *Football Math: Touchdown Activities and Projects for Grades 4–8* published by Good Year Books. Copyright © 1995, 2005 Jack Coffland and David A. Coffland.

4. The positions of quarterback, running back, receiver, and tight end have been called the "skill positions." What is the cost for an average team for the skill positions?

_____

5. What is the average salary for an offensive team? For a defensive team? Why is one higher than the other?

_____

Average Quarterback Salaries

Average Quarterback Salaries

Player	Above Average	Below Average
Payton Manning		
Donovan McNabb		
Chad Pennington		
Rex Grossman		
Tim Rattay		
Tommy Maddox		

1. Here are the names of six NFL quarterbacks. Do you think their salaries would be above or below the average given for the quarterbacks? Write in your prediction for each player. Add a statement as to why you think the salaries would be above or below average.

_____

_____

_____

_____

_____

_____

_____

_____

_____

_____

2. Why do you think quarterbacks are the highest-paid players?

_____

3. Now that you've examined quarterbacks who are above or below the average salary, look again at the chart on Page 60. Why do you think punters are the lowest-paid players?

_____

4. Using the chart on Page 60, list other players at other positions and estimate if they are paid above or below the league averages.

_____

_____

_____

Quarterback Statistics

When we examine statistics, it is sometimes easy to jump to conclusions. For example, when you examine the statistical chart, you find that Steve McNair threw the fewest interceptions, right? But he also threw the fewest passes. Is that important? Yes. One way to compare statistics is to change them to percentages so we get a comparison based upon 100. Then we are saying, "If they all threw 100 passes, what number would have been intercepted?" Using this number we can compare the quarterbacks in a better manner.

Leading Passers, 2003

Player & Team	Passes Attempted	Passes Completed	TDs	Interceptions
Payton Manning, IND	566	379	29	10
Donovan McNabb, PHL	478	275	16	11
Steve McNair, TN	400	250	24	7
Dante Culpepper, MN	454	295	25	11

1. Compute the percentage completion figure for each quarterback. (Do this by comparing the number of completions and the number of passes attempted.) Then rate them from top to bottom by that percentage.

   _____

2. Compute the percentage of interceptions to passes attempted for each quarterback. Then rate them from top to bottom by that percentage.

   _____

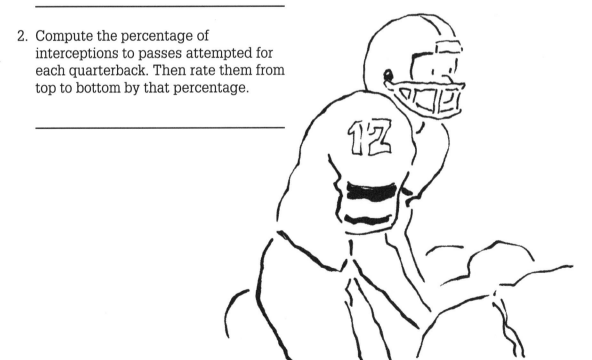

Quarterback Statistics (cont'd.)

Solve the following problems.

3. Compute the percentage of touchdowns to passes attempted for each quarterback. Then rate them from top to bottom by that percentage.

_____

_____

_____

_____

4. Now enter your numbers into the chart below.

Percentage of Passes That Were:

Player & Team	Completed	Intercepted	Touchdowns
Payton Manning, IND			
Donovan McNabb, PHL			
Steve McNair, TN			
Dante Culpepper, MN			

5. Which quarterback do you think played the best? Why? Is there a right answer to this question?

_____

_____

_____

_____

From *Football Math: Touchdown Activities and Projects for Grades 4–8* published by Good Year Books. Copyright © 1995, 2005 Jack Coffland and David A. Coffland.

The Price of Victory

You are the business manager for your high school football team. You have won all your games to this point. It is now time to plan the trip to the state championship game. The athletic association has promised you they will help meet the expenses. Plan your budget. How much will you have to spend?

Note: You must provide for the expenses of 45 football players, six coaches, and three team managers. You must stay at a hotel for two nights before the game and one night after the game.

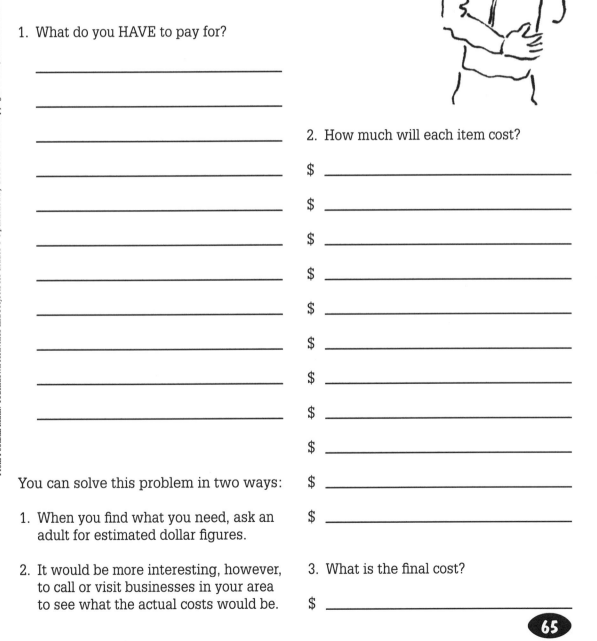

1. What do you HAVE to pay for?

_____

_____

_____

_____

_____

_____

_____

_____

_____

You can solve this problem in two ways:

1. When you find what you need, ask an adult for estimated dollar figures.

2. It would be more interesting, however, to call or visit businesses in your area to see what the actual costs would be.

2. How much will each item cost?

$ _____

$ _____

$ _____

$ _____

$ _____

$ _____

$ _____

$ _____

$ _____

$ _____

$ _____

3. What is the final cost?

$ _____

Here Comes the Turk

An NFL team can keep 45 players on its active roster. That is not very many, when you consider that a coach must have offensive players, defensive players, and kicking specialists. Listed below are the positions that must be filled on an NFL team. (For simplicity, we have assigned the defensive team to play four down linemen and three linebackers.)

Offensive Positions	Defensive Positions
Split End	Left Defensive End
Left Tackle	Left Defensive Tackle
Left Guard	Right Defensive Tackle
Center	Right Defensive End
Right Guard	Left Outside Linebacker
Right Tackle	Middle Linebacker
Tight End	Right Outside Linebacker
Quarterback	Left Corner
Running Back	Free Safety
Fullback	Strong Safety
Wide Receiver	Right Corner
Place Kicker	Punter

How many players should you keep for each position? Remember, you have a 45-man limit on the roster. Players do get hurt; you will need to have substitutes for every position. However, a substitute might play more than one position.

Also, consider the following in making your choices:

a. Coaches like to keep three quarterbacks. If one gets hurt, it is difficult to hire a quarterback and teach him the offense in the middle of the year.

b. Coaches like to keep three tight ends. This gives them extra blockers for goal line plays.

From *Football Math: Touchdown Activities and Projects for Grades 4–8* published by Good Year Books. Copyright © 1995, 2005 Jack Coffland and David A. Coffland.

c. Coaches like to have extra defensive backs to use in the "nickel" and "dime" packages. (These terms describe having five or six defensive backs in the game playing for obvious passing situations.)

d. No NFL team uses the same player for punting and place kicking.

1. During the summer, when teams are trying to cut players to get to the 45-player limit, make a list of all the players on your favorite team (returning players, newly signed players, and new draft choices). See if you can "help the coach" get down to the 45-player limit.

2. Watch for the new "salary cap." Use the salary information in these projects to see if your team can afford the players you select.

Weight-lifting Excercises

Football players lift weights to build up their muscles and their weight. Over a season of lifting weights, the average player may move enormous amounts. Help "Walt the Weight Lifter" figure out what he lifts in a year.

1. Walt lifts weights all 52 weeks of the year.

2. Walt lifts weights 4 times a week.

3. Walt performs four lifts:
   a. Bench press   210 pounds
   b. Incline press   170 pounds
   c. Squats   380 pounds
   d. Curls   80 pounds

4. Walt lifts the weights shown above each time for each lift.

5. Walt performs each lift 20 times.

Using the weights shown above for each of Walt's lifts:

a. How much weight does Walt move in a year for each lift?

_____

b. How much total weight does Walt move in a year with all of the lifts?

_____

c. How many weeks will it take Walt to join the "Million Pound Club" for all of his lifts put together? (The "Million Pound Club" is an elite group of lifters who have all lifted at least one million pounds.)

_____

d. How many weeks will it take Walt to join the "Million Pound Club" for each lift?

_____

Because the numbers will be large, you may want to use a calculator to help you with the computations. Also, ask your coach, or read in a magazine, which muscles each of the weight lifts builds.

From *Football Math: Touchdown Activities and Projects for Grades 4–8* published by Good Year Books. Copyright © 1995, 2005 Jack Coffland and David A. Coffland.

Weight-lifting Sets

Tommy the Big Tackle, who plays for the Cougars, lifts weights in the off-season to keep his muscles strong. He follows the plan listed below.

Tommy does not want to hurt himself, so he practices with less than the total amount of weight he can lift. His practice schedule is shown below:

Set 1: He lifts 60% of his maximum load ten times. He repeats this set three times, with a short pause between each set of ten lifts.

Set 2: He lifts 70% of his maximum load six times. He repeats this set two times, with a short pause between each set of six lifts.

Set 3: He lifts 80% of his maximum load two times. This is done only once.

Tommy does each of these sets with his four different lifts. His lifts, and his maximum weight lifted for each lift, is shown below:

    a. Bench Press   240 pounds
    b. Incline Press   210 pounds
    c. Squats   410 pounds
    d. Curls   90 pounds

1. How much weight does Tommy practice with for Set 1? Set 2? Set 3?

_____

2. How much total weight does Tommy lift in Set 1? Set 2? Set 3?

_____

3. If Tommy lifts weights 6 days a week, how much weight does he lift in a week?

_____

4. All together, Tommy lifts weights 30 weeks of the year. How much total weight does he lift?

_____

5. How long will it take Tommy to join the "Million Pound Club?"

_____

From Football Math: Touchdown Activities and Projects for Grades 4–8 published by Good Year Books. Copyright © 1995, 2005 Jack Coffland and David A. Coffland.

Frequent Flyer Miles

Football teams travel an amazing number of miles over the years as they fly to play games. Consider the regular-year schedule for two teams, one located in the center of the country and one located in a corner of the country.

Sample Schedule, 2004 NFL Season

	Chicago Bears		Seattle Seahawks	
Week 1	Detroit	Home	New Orleans	Away
Week 2	Green Bay	Away	Tampa Bay	Away
Week 3	Minnesota	Away	San Francisco	Home
Week 4	Philadelphia	Home	Bye	
Week 5	Bye		St. Louis	Home
Week 6	Washington	Home	New England	Away
Week 7	Tampa Bay	Away	Arizona	Away
Week 8	San Francisco	Home	Carolina	Home
Week 9	N.Y. Giants	Away	San Francisco	Away
Week 10	Tennessee	Away	St. Louis	Away
Week 11	Indianapolis	Home	Miami	Home
Week 12	Dallas	Away	Buffalo	Home
Week 13	Minnesota	Home	Dallas	Home
Week 14	Jacksonville	Away	Minnesota	Away
Week 15	Houston	Home	N.Y. Jets	Away
Week 16	Detroit	Away	Arizona	Home
Week 17	Green Bay	Home	Atlanta	Home

From *Football Math: Touchdown Activities and Projects for Grades 4–8* published by Good Year Books. Copyright © 1995, 2005 Jack Coffland and David A. Coffland.

Frequent Flyer Miles (cont'd.)

As you can see, each team plays eight home games and eight away games during the season, and each has two weeks in which they do not play. The question is, how far does each fly? The away sites are listed in the chart below, with the one-way distance between Chicago/Seattle and their opponents.

Chicago flies to: City/Team	Mileage	Seattle flies to: City/Team	Mileage
Green Bay	156	New Orleans	2,098
Minnesota	304	Tampa Bay	2,526
Tampa Bay	1,004	New England	2,488
N.Y. Giants	714	Arizona	1,112
Tennessee	397	San Francisco	684
Dallas	802	St. Louis	1,717
Jacksonville	874	Minnesota	1,382
Detroit	196	N.Y. Jets	2,404

1. How far does each team fly during a season? (Remember, the distances shown are one-way mileages.)

_____

2. Who flies further?

_____

3. Figure out the time zone changes that each team has to go through on each trip. Who do you think has the easier travel schedule? Why?

_____

_____

_____

_____

Tampa Bay vs. Green Bay Records

Comparing team records is not always fair because some teams have been in the league longer than others. Here are some team records for the Green Bay Packers, a long-time NFL team, and the Tampa Bay Buccaneers, a team that joined the NFL in the 1970s. Compare the records

Green Bay Records

Category	Player, Year	Record
Career Rushing Yards	Jim Taylor, 1958–66	8,207 yds.
Career Passing Yards	Brett Favre, 1992–2003	45,646 yds.[†]
Career Receiving Yards	James Lofton, 1976–86	9,656 yds.
Career Points	Ryan Longwell, 1997–2003	844 pts.[†]
Season Rushing	Ahman Green, 2003	1,833 yds.
Season Passing	Lynn Dickey, 1983	4,458 yds.
Season Receiving	Robert Brooks, 1995	1,497 yds.
Season Points	Paul Hornung, 1960	176 pts.[*]
Game Rushing	Ahman Green, 12/28/2003	190 yds.
Game Passing	Lynn Dickey, 10/12/1980	418 yds.
Game Receiving	Bill Howton, 10/21/1956	257 yds.
Game Points	Paul Hornung, 10/8/1961	33 pts.

*NFL Record

†Through the 2003 season

Tampa Bay vs. Green Bay Records (cont'd.)

Tampa Bay Records

Category	Player, Year	Record
Career Rushing Yards	James Wilder, 1981–89	5,957 yds.
Career Passing Yards	Vinny Testaverde, 1987–92	14,820 yds.
Career Receiving Yards	Mark Carrier, 1987–92	5,018 yds.
Career Points	Martin Gramatica, 1999–2003	538 pts.*
Season Rushing	James Wilder, 1984	1,544 yds.
Season Passing	Brad Johnson, 2003	3,811 yds.
Season Receiving	Mark Carrier, 1989	1,422 yds.
Season Points	Martin Gramatica, 2002	128 pts.
Game Rushing	James Wilder, 11/6/1983	219 yds.
Game Passing	Doug Williams, 11/16/1980	486 yds.
Game Receiving	Mark Carrier, 12/6/1987	212 yds.
Game Points	Jimmie Giles, 10/20/1985	24 pts.

*Through the 2003 season

Compare the records for each team. What do you see?

1. Are the Packers' career records higher? What are the dates of their records? Was Tampa Bay even in the league during those years?

2. Are the Packers' season records higher? Are the Packers' individual game records higher?

3. Only one record for these two teams is an NFL record. Why do you think Paul Hornung's record for points scored in a season was so high? (*Hint:* He was a runner *and* a kicker.)

Second-place Seasons

Buffalo and Minnesota are two teams with interesting Super Bowl records. They had good teams for several years; they each played in four different Super Bowls. However, both teams lost all four of their games. Here are the statistics about those games.

Minnesota and Buffalo in the Super Bowl

Bowl	Winner	Score	Opponent	Score
SB IV	Kansas City	23	Minnesota	7
SB VIII	Miami	24	Minnesota	7
SB IX	Pittsburgh	16	Minnesota	6
SB XI	Oakland	32	Minnesota	14
SB XXV	N.Y. Giants	20	Buffalo	19
SB XXVI	Washington	37	Buffalo	24
SB XXVII	Dallas	52	Buffalo	17
SB XXVIII	Dallas	30	Buffalo	13

Go to a library to look for information on these games. You may find the information in old newspapers, almanacs, or magazines. Or you can look on the Internet for sites about the Super Bowl and its history. Then answer the following questions:

1. Both of these teams had famous quarterbacks during some of their games. Who were these quarterbacks?

   _____

   _____

2. Look at the game statistics. Were the games as lopsided as their scores?

   _____

a. Did Buffalo or Minnesota have a large number of turnovers that gave away points? In which games?

   _____

   _____

   _____

b. Were Buffalo or Minnesota ever ahead in one or more of the games? Has Minnesota ever been ahead in a Super Bowl? Has Minnesota ever scored in the first half of a Super Bowl?

_____

_____

_____

_____

c. What else can you find out about these games?

_____

_____

_____

_____

Note: Denver also used to have an 0–4 record in Super Bowls until they won back-to-back Super Bowls in 1998 and 1999. Can you find records for all of Denver's losses?

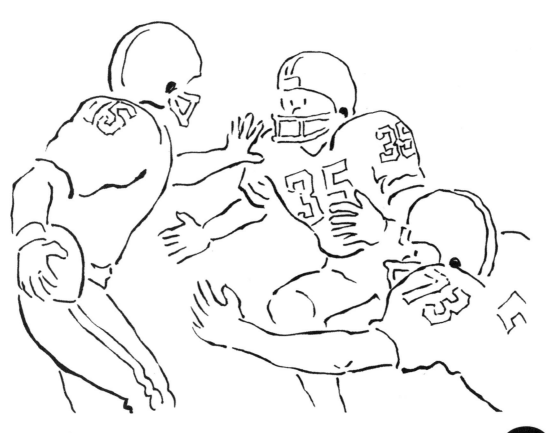

Comparing Passing Teams

During the early years of this century, New England has gained a reputation as an effective passing team. Here are some statistics for New England during 2003, a season that saw them win the Super Bowl.

Leading New England Rushers, 2003

Player	Carries	Yards Gained	Longest	TDs
Team	*473*	*1,607*	*42*	*9*
Antowain Smith, RB	182	642	30	3
Kevin Faulk, RB	178	638	23	0
Mike Cloud, RB	27	118	42	5
Larry Centers, RB	21	82	13	0
Tom Brady, QB	42	63	11	1

Leading New England Passers, 2003

Player	Attempts	Completions	Yards	TDs
Team	*537*	*320*	*3,432*	*23*
Tom Brady, QB	527	317	3,620	23
Rohan Davey, QB	7	3	31	0

1. New England did not have a 1,000-yard rusher in 2003. But did the team have two rushers who totaled more than 1,000 yards? Who were they? Who gained more, and by how much? What do these rushing totals say about the way New England ran the ball?

   _____

   _____

   _____

From *Football Math: Touchdown Activities and Projects for Grades 4–8* published by Good Year Books. Copyright © 1995, 2005 Jack Coffland and David A. Coffland.

2. Check the number of passing attempts and rushing attempts for New England. Now do the same for San Diego on the next page. Do these figures explain why one team is thought of as a passing team while the other is considered a rushing team?

_____

Leading New England Receivers, 2003

Player	Receptions	Yards Gained	Longest	TDs
Team	*320*	*3,432*	*82*	*23*
Deion Branch, WR	57	803	66	3
David Givens, WR	34	510	57	6
Troy Brown, WR	40	472	82	4
Kevin Faulk, RB	48	440	27	0
Daniel Graham, TE	38	409	38	4
Christian Fauria, TE	28	285	28	2

3. Check the number of passing yards gained and rushing yards gained for both teams. Do these figures explain why one team is famous for passing and the other for running?

_____

Comparing Running Teams

During the time New England has gained a reputation as an effective passing team, San Diego has become a rushing team with a star running back. Here are some the statistics for San Diego during 2003, a season in which they won 4 and lost 12 games.

Leading San Diego Rushers, 2003

Player	Carries	Yards Gained	Longest	TDs
Team	*417*	*2,146*	*73*	*16*
LaDainian Tomlinson, RB	313	1,645	73	13
Doug Flutie, QB	33	168	17	2
Tim Dwight, WR	9	88	20	0
Drew Brees, QB	21	84	18	0
Lorenzo Neal, RB	18	40	7	1

Leading San DiegoPassers, 2003

Player	Attempts	Completions	Yards	TDs
Team	*525*	*297*	*3,021*	*21*
Drew Brees, QB	356	205	2,108	11
Doug Flutie, QB	167	91	1,097	9

From *Football Math: Touchdown Activities and Projects for Grades 4–8* published by Good Year Books. Copyright © 1995, 2005 Jack Coffland and David A. Coffland.

Comparing Running Teams (cont'd.)

Leading San Diego Receivers, 2003

Player	Receptions	Yards Gained	Longest	TDs
Team	*297*	*3,021*	*73*	*21*
David Boston, WR	70	880	46	7
LaDainian Tomlinson, RB	100	725	73	4
Antonio Gates, TE	24	389	48	2
Kassim Osgood, WR	13	278	57	2
Eric Parker, WR	18	244	33	3
Tim Dwight, WR	14	193	32	0

1. Find the total number of offensive plays for New England and San Diego. Then: What percentage of the offensive plays were runs for both teams? What percentage of the offensive plays were passes for both teams?

   _____

   _____

2. San Diego's star running back, LaDainian Tomlinson, caught more passes than any of his team's wide receivers. How many passes did San Diego's wide receivers catch?

   _____

   _____

3. Find the percentage of both teams' receptions caught by running backs listed on the chart. Now find the percentage of receptions caught by both teams' wide receivers. Do these data suggest a reason one team is considered a passing team and the other a running team?

   _____

   _____

Professional Statistics

Select your favorite professional football team. Then try to keep track of the statistics for the team for an entire season. You may

1. Keep team statistics. Look at the statistical information presented in your paper for each game and decide what team statistics you would like to keep. Consider things such as the following:

2. Keep the same statistics for your opponent each week.

Offensive Statistics	Defensive Statistics	Combined Statistics
Points scored	Fumbles caused	Total penalties
Total first downs	Fumbles recovered	Penalty yardage
Running yardage	Passes intercepted	Kick-off return yardage
Total number of running plays	Interception return yardage	Punt return yardage
Passing yardage	Tackles	
Passes attempted	Sacks	
Passes completed		
Passes had intercepted		
Total fumbles		
Fumbles lost		

Then, you can figure out:

3. What were the average figures for each of the games during the year?

_____

_____

_____

What were the average figures for each win? For each loss?

_____

_____

For each average statistic over the year, did your favorite team do better than the opponent? Where did they do better? Where did they do worse?

_____

_____

_____

4. Did your favorite team win more games or lose more games? From your statistics, can you explain why?

_____

_____

_____

College Statistics

Select your favorite college football team. Then try to keep track of the statistics for the team for an entire season. You may want to consider the following:

1. Keep team statistics. Look at the statistical information presented in your paper for each game, and decide what team statistics you would like to keep. Consider things such as the following:

2. Keep the same statistics for your opponent each week.

Offensive Statistics	Defensive Statistics	Combined Statistics
Points scored	Fumbles caused	Total penalties
Total first downs	Fumbles recovered	Penalty yardage
Running yardage	Passes intercepted	Kick-off return yardage
Total number of running plays	Interception return yardage	Punt return yardage
Passing yardage	Tackles	
Passes attempted	Sacks	
Passes completed		
Passes had intercepted		
Total fumbles		
Fumbles lost		

From *Football Math: Touchdown Activities and Projects for Grades 4–8* published by Good Year Books. Copyright © 1995, 2005 Jack Coffland and David A. Coffland.

From *Football Math: Touchdown Activities and Projects for Grades 4–8* published by Good Year Books. Copyright © 1995, 2005 Jack Coffland and David A. Coffland.

Then you can figure out:

3. What were the average figures for each of the games during the year?

_____

_____

_____

What were the average figures for each win? For each loss?

_____

_____

For each average statistic over the year, did your favorite team do better than the opponent? Where did they do better? Where did they do worse?

_____

_____

_____

4. Did your favorite team with more games or lose more games? From your statistics, can you explain why?

_____

_____

_____

Comparing Statistics

Now that you have statistics for your favorite professional and college football teams, compare them. Then answer some of the following questions.

1. What is the average number of offensive plays for a professional team? For a college team?

_____

_____

2. What is the average number of points scored for both teams?

_____

_____

3. What is the average number of running plays and passing plays for a college team? For a professional team?

_____

_____

4. Can you find any relationships here? Does one team run more plays than the other? Do they score more points if they run more plays?

_____

_____

5. Who scores more of their points on touchdowns? Who scores more of their points on field goals? What kind of numbers will you have to work with to answer these questions?

_____

_____

6. What other questions can you ask and answer from the statistics you have compiled?

_____

_____

College Quarterback Stats

On pages 82–83, you learned how to figure the college passer's rating statistics. Let's review it one more time.

You need the following statistics from each game:

> Pass completions
> Pass attempts
> Yards gained passing
> Number of passing touchdowns
> Number of interceptions thrown

Then, make the following computations:

Step 1. Divide the pass completions by the pass attempts; multiply the answer by 100.

Step 2. Divide the yards gained passing by the pass attempts; multiply the answer by 8.4.

Step 3. Divide passing touchdowns by attempts; multiply the answer by 100 and then 3.3.

Step 4. Divide interceptions by attempts; multiply the answer by 100 and then by 2.

Step 5. Add answers from Steps 1, 2, and 3; then subtract the answer from step 4.

College Quarterback Stats (cont'd.)

Pick your favorite college quarterback. Find his statistics for each game during an entire season. Compute the statistic for each game, but compute it each week for his season totals. At the end of the year, you will have his passing rating for the entire year.

Suggestions:

1. Have some friends pick their favorite quarterback and compute his statistics.

2. Compare the passing ratings for each game. Whose favorite had the best day?

3. Compare the total passing rating score each week for your quarterbacks. Who is having the best season?

4. At the end of the year, compare your quarterbacks. Who had the highest final rating? Award that quarterback your own "Most Valuable Passer" award.

Who Is the Best Kicker?

If a team cannot score a touchdown, then it is vital that they make a field goal if they get close to the other team's goal. It is almost like "gift points" when a team scores a 50± yard field goal, as it means the team scored points when it wasn't close to the other team's goal line.

Kicking Data for Selected Kickers, 2003–04

| Player, Team | Points | Kicks Made–Kicks Attempted | | | | | Longest Kick |
		1–19 Yards	20–29 Yards	30–39 Yards	40–49 Yards	50–59 Yards	
Jeff Wilkins, STL	163	0–0	16–16	11–13	8–9	4–4	53
Mike Vanderjagt, IND	157	0–0	17–17	7–7	12–12	1–1	50
Jason Elam, DEN	120	0–0	10–11	6–6	9–11	2–3	51
Adam Vinatieri, NE	112	0–0	16–17	4–8	5–8	0–1	48

The chart on this page gives you some interesting statistics on four kickers who were among the leading scorers for kickers during the 2003–04 season. Examine the figures. Then pair off with a friend. Discuss these kickers in terms of the following:

1. Who was the best kicker from close range?

   _____

2. Who was the best kicker from long range?

   _____

3. What do you notice about Mike Vanderjagt's statistics?

   _____

Who Is the Best Kicker? (cont'd.)

4. Who scored the most points on field goals? (Remember the chart gives you total points, not field goal points.)

_____

5. Which kicker would you rather have if you were a coach? Why?

_____

_____

6. Who do you think is the best kicker? Is it the same person that you would like to have on your team? Why or why not?

_____

_____

_____

_____

_____

From *Football Math: Touchdown Activities and Projects for Grades 4–8* published by Good Year Books. Copyright © 1995, 2005 Jack Coffland and David A. Coffland.

Kicking Statistics for the Season

Collect as many statistics as you can on several kickers who are kicking this year. Keep the statistics for an entire year. (You may want to keep statistics for several kickers, as it is always possible that someone will be hurt and not kick very much during a year.) Consider keeping the following statistics:

a. Number of kickoffs
b. Number of kickoffs kicked into the end zone for a touchback
c. Number of kickoffs returned for a touchdown
d. Average length of kickoff returns (This is important: Kickers who kick the ball high usually cause the other team to make shorter returns.)
e. Number of field goals attempted and made
f. Number of point after touchdowns attempted and made

g. Number of field goals blocked
h Number of points after touchdowns blocked
i. Length of all field goal kicks attempted and made
j. Any other statistic you think is important

Then show your statistics to a friend and discuss who is the best kicker.

The "Games Behind" Statistic

The table below shows the team standings for an imaginary football league. the interesting statistics are in the "Games Behind" column. Let's examine it.

The Bruins are in the lead; therefore, the statistic is not figured for them. The Cardinals are in second place. Notice they have lost two more games and won two fewer games than the Bruins. As a result, if the Bruins lose their next two games and the Cardinals win their next two, the two teams will be tied. The "Games Behind" statistic tells you this number. It tells the number of Cardinal wins and Bruin losses that would result in a tie for first place.

Pioneer League Football Standings

Team	Wins	Losses	Ties	Games Behind
Bruins	8	1	0	
Cardinals	6	3	0	2
Lions	5	4	0	3
Jets	4	5	0	4
Dogs	0	9	0	8

The first table is easy to read, because everyone has played the same number of games. That is not always true. Consider the next set of standings:

From *Football Math: Touchdown Activities and Projects for Grades 4–8* published by Good Year Books. Copyright © 1995, 2005 Jack Coffland and David A. Coffland.

The "Games Behind" Statistic (cont'd.)

Pioneer League Football Standings

Team	Wins	Losses	Ties	Games Behind
Bruins	8	0	0	
Cardinals	6	1	1	1-1/2
Lions	5	4	0	3-1.5
Jets	4	4	0	4
Dogs	2	5	1	5-1/2

This set of standings happens more often. Everyone has not played the same number of games. Two teams have played a tie game. How are those things figured?

Again, look at the Bruins and the Cardinals. The Bruins are in first place; the statistic is not figured for them. The Cardinals have won two fewer games than the Bruins, but they have lost 1 more. This puts the Cardinals 1-1/2 games behind the Bruins. They are two games behind in the Wins column, but only one game behind in the Losses column. (The tie does not help or hurt in figuring this statistic. However, at the end of the season when everyone has played the same number of games, it makes a major difference in who has the most victories or the most losses.)

Follow your favorite team or teams for an entire season. Read the league standings in any newspaper. Be prepared to discuss how your team can win or lose the league title.

Analyzing Records from AFC History

Part of "enjoying football" is keeping track of statistics and discussing them with other football fans. But when you examine statistics, you must be aware of the changes that can impact the data.

Records of AFC Teams from 1970 through 1992

Team	Wins	Losses	Ties	Pct.	Div. Titles	Play-off Berths	Post-season Record	Super Bowl Record
Miami Dolphins	229	113	2	.669	10	14	16–12	2–3
L.A. Raiders	217	121	6	.641	9	14	17–11	3–0
Pittsburgh Steelers	203	140	1	.592	10	13	16–9	4–0
Denver Broncos	193	140	6	.570	7	9	9–9	0–4
Cleveland Browns	171	170	3	.501	6	9	3–9	0–0
Cincinnati Bengals	172	172	0	.500	5	7	5–7	0–2
Kansas City Chiefs	157	180	7	.466	1	5	5–7	0–0
Seattle Seahawks	121	139	0	.465	1	4	3–4	0–0
Buffalo Bills	157	185	2	.459	5	8	9–8	0–3
San Diego Chargers	151	188	5	.446	4	5	4–5	0–0
New England Patriots	153	191	0	.445	2	5	3–5	0–1
Houston Oilers	152	190	2	.444	1	9	7–9	0–0
N.Y. Jets	144	198	2	.422	0	5	3–5	0–0
Indianapolis Colts	144	198	2	.422	5	6	4–5	1–0

So, the chart shows the statistical records for all AFC teams from 1970 (the official year of the merger of the American Football Conference with the National Football Conference) to 1992. What are the changes that might make some of the statistics appear incorrect?

_____

_____

1. There were Super Bowls played before 1970. So, for example, the early Super Bowl records of the Kansas City Chiefs and the New York Jets do not show in the Super Bowl Record column.

Can you find the Super Bowls that were played in 1970 or before and add those wins and losses to the records of the correct teams? _Hint:_ You should be able to find four additional Super Bowl games.

2. The teams playing in the AFC are not the same. Some teams have moved to new locations; others have changed to or from the AFC. Can you find the history of your team? For example, four teams listed in this chart have changed in some way: the LA Raiders, the Cleveland Browns, the Houston Oilers, and the Seattle Seahawks. Which team:

a. Moved to become the Tennessee Titans?

_____

b. Moved back to the city where the team began years ago?

_____

c. Moved from the AFC to the NFC in 2002?

_____

d. Moved from the city shown in the chart? A new, expansion team was given to the original city several years later.

_____

3. It is almost impossible to update this chart because of all the changes in NFL conferences, divisions, and teams. But if you want to discuss records, what can you do? The NFL reorganized all of its divisions after the 2001 season. First, update the chart for these the Super Bowls from 1993 to 2001. Next, find the records for the 2002 and 2003 seasons and begin a new chart that shows the AFC as it exists now. Can you add new seasons (2004 or later) to your chart as each season is finished?

4. After adding the Super Bowls played before 1970 and after 1992, name the AFC teams that have yet to win a Super Bowl. Which have yet to even play in a Super Bowl?

_____

5. Has anyone played in a tie game since 1992? (Remember, the NFL added a tie-breaking system, but if no team has scored after one additional period in a regular-season game, the game can end in a tie. That is not true for a play-off game; if no team has won after one additional period, they play another and another until someone wins finally wins!)

_____

6. What is your favorite AFC team? Can you update their record from 1992 to the present?

_____

Analyzing Records from NFC History

Part of "enjoying football" is keeping track of statistics and discussing them with other football fans. But when you examine statistics, you must be aware of the changes that can impact the data.

Records of NFC Teams from 1970 through 1992

Team	Wins	Losses	Ties	Pct.	Div. Titles	Play-off Berths	Post-season Record	Super Bowl Record
Washington Redskins	221	122	1	.644	5	13	18–10	3–2
Dallas Cowboys	215	129	0	.625	10	16	23–13	3–3
San Francisco 49ers	202	139	3	.592	12	13	17–9	4–0
Minnesota Vikings	201	141	2	.587	11	14	11–14	0–3
L.A. Rams	198	142	4	.582	8	14	10–14	0–1
Chicago Bears	179	164	1	.522	6	9	6–8	1–0
Philadelphis Eagles	163	175	6	.482	2	8	4–8	0–1
N.Y. Giants	158	184	2	.462	3	6	9–4	2–0
Detroit Lions	150	190	4	.442	2	4	1–4	0–0
Phoenix Cardinals	144	194	6	.427	2	3	0–3	0–0
Green Bay Packers	141	195	8	.421	1	2	1–2	0–0
New Orleans Saints	140	200	4	.412	1	4	0–4	0–0
Atlanta Falcons	138	202	4	.406	1	4	2–4	0–0
Tampa Bay Buccaneers	76	183	1	.294	2	3	1–3	0–0

From *Football Math: Touchdown Activities and Projects for Grades 4–8* published by Good Year Books. Copyright © 1995, 2005 Jack Coffland and David A. Coffland.

So, the chart shows the statistical records for all NFC teams from 1970 (the official year of the merger of the American Football Conference with the National Football Conference) to 1992. What are the changes that might make some of the statistics appear incorrect?

_____

_____

1. There were Super Bowls played before 1970. So, for example, the early Super Bowl wins of the Green Bay Packers do not show in the Super Bowl Record column.

   Can you find the Super Bowls that were played in 1970 or before and add those wins and losses to the records of the correct teams? *Hint:* You should be able to find four additional Super Bowl games.

From *Football Math: Touchdown Activities and Projects for Grades 4–8* published by Good Year Books. Copyright © 1995, 2005 Jack Coffland and David A. Coffland.

2. The teams playing in the NFC are not the same. Some teams have moved to new locations; others have changed to or from the NFC. Can you find the history of your team? For example, two teams listed in this chart have changed in some way: the LA Rams and the Phoenix Cardinals. In addition, there is a new team in the NFC, and a team has moved from the AFC to the NFC. Which team:

- Moved to St. Louis?

  _____

- Changed its name to represent the state, not the city?

  _____

- Is not on this list, but moved to the NFC from the AFC?

  _____

- Is not on this list, because it is an NFC expansion team formed in 1999?

  _____

3. It is almost impossible to update this chart because of all the changes in NFL conferences, divisions, and teams. But if you want to discuss records, what can you do?

  _____

The NFL reorganized all of its divisions after the 2001 season. First, update the chart for these the Super Bowls from 1993 to 2001. Next, find the records for the 2002 and 2003 seasons and begin a new chart that shows the NFC as it exists now. Can you add new seasons (2004 or later) to your chart as each season is finished?

4. After adding the Super Bowls played before 1970 and after 1992, name the NFC teams that have yet to win a Super Bowl. Which have yet to even play in a Super Bowl?

  _____

5. Has anyone played in a tie game since 1992? (Remember, the NFL added a tie-breaking system, but if no team has scored after one additional period in a regular-season game, the game can end in a tie. That is not true for a play-off game; if no team has won after one additional period, they play another and another until someone wins finally wins!)

  _____

6. What is your favorite NFC team? Can you update their record from 1992 to the present?

  _____

All-purpose Yards Leaders II

All-purpose yardage is a statistic that combines rushing and receiving yards with yardage from all runbacks. This statistic gives credit to the players who "do it all" on the field. Athletes who do well in this statistical category have a variety of skills that make them dangerous to opposing defenses. The NCAA record holder for most all-purpose yards in one season is Barry Sanders of Oklahoma State University. During the 1988 season, he rushed for 2,628 yards, had 106 receiving yards, returned punts for 95 yards, and added 421 more yards on kickoff returns, for a total of 3,250 all-purpose yards.

You can use the following chart to track the all-purpose yards of your favorite player.

All-purpose Yards Leaders

Opponent	Rushing Yards	Receiving Yards	Punt Return Yards	Kickoff Return Yards	All-purpose Yards
Season Totals:					

Fill in the opposing team's name for each game. Then write the player's rushing, receiving, and return yards in the appropriate column. Add up all the numbers in each row to find the all-purpose statistic for each game. At the end of the season, add up all the columns to find the season totals. In the lower, right-hand corner of the chart, fill in the player's all-purpose yards for the season.

From *Football Math: Touchdown Activities and Projects for Grades 4–8* published by Good Year Books. Copyright © 1995, 2005 Jack Coffland and David A. Coffland.

Miami's Decade of Champs

The University of Miami had an excellent record for the 10 seasons from 1983–92. Their record looks like this:

All-purpose Yards Leaders

Season	Wins	Losses	Ties	National Ranking
1983	11	1	0	1st—National Champs
1984	8	5	0	18
1985	10	2	0	9
1986	11	1	0	2
1987	12	0	0	1st—National Champs
1988	11	1	0	2
1989	11	1	0	1st—National Champs
1990	10	2	0	3
1991	12	0	0	1st—National Champs
1992	11	1	0	3

Questions to consider:

1. How many total wins did Miami have during the decade?

   _____

2. How many total losses did they have?

   _____

3. What was their winning percentage for the 10 years?

   _____

4. What other statistics might we collect?

   _____

From *Football Math: Touchdown Activities and Projects for Grades 4–8* published by Good Year Books. Copyright © 1995, 2005 Jack Coffland and David A. Coffland.

5. How many points did the team score each year?

_____

6. How many opponents' points did they allow?

_____

7. How may TDs?

_____

8. How many field goals?

_____

9. How many safetys?

_____

Collect the same record for your favorite team. You might select a high school team, a college team, or an NFL team. The sports desk of your local newspaper can help you find the records for past seasons. Or, for college or professional teams, buy statistical record books that list the information.

Comparing Your Favorite Teams

It is fun to compare records of your favorite teams with a friend. If you have different favorite teams, you can talk about who has the best record or who is improving the most. Consider the following 10-year record for two professional teams, one from the American Football Conference and one from the National Football Conference.

	AFC: Denver Broncos		NFC: Minnesota Vikings	
Year	Wins	Losses	Wins	Losses
2003	10	6	9	7
2002	9	7	6	10
2001	8	8	5	11
2000	11	5	11	5
1999	6	10	10	6
1998	14	2	15	1
1997	12	4	9	7
1996	13	3	13	3
1995	8	8	8	8
1994	7	9	10	6

You might wish to consider:

1. Who has the best 10-year record?

   _____

2. Who has the best 5-year record?

   _____

3. What is the total number of wins for each team?

   _____

4. What is the winning percentage for each team?

   _____

Comparing Your Favorite Teams (cont'd.)

You might like to discuss:

5. Who seems to be getting better? Which team's record seems to be getting worse?

_____

_____

6. What happens if you add 1993 statistics? How do the records change?

_____

_____

7. Can you add statistics from any later year? How do they affect the records?

_____

_____

From *Football Math: Touchdown Activities and Projects for Grades 4–8* published by Good Year Books. Copyright © 1995, 2005 Jack Coffland and David A. Coffland.

Comparing Teams and Travel

On the following pages you will find these mileage charts: ,

a. AFC–NFC miles: The miles between NFC cities and AFC cities.
b. National Football Conference miles: The miles between each NFC city.
c. American Football Conference miles: The miles between each AFC city.

Use the charts for the following:

1. Get this year's schedule for your favorite team. How far do they have to travel this year?

   _____

2. Can you compare this year's travel to last year's travel schedule? Do they travel more miles? Fewer miles?

   _____

3. Have a friend figure the miles his or her favorite team travels. Compare that to your team. Who has the more difficult travel schedule? Why?

   _____

   _____

4. Look at the time zone changes for each trip. How many times does your favorite team cross three time zones? Two time zones? One time zone?

No time zones? Make a chart that shows this information. Compare your chart to a chart showing the same information for your friend's team. Do you think it makes it more difficult to play when a team has to change three time zones?

   _____

   _____

Compare all the trips from all the charts:

5. What is the longest trip any team has to make?

   _____

6. What is the shortest trip any team has to make?

   _____

Comparing Teams and Travel (cont'd.)

6. Compare your favorite team's winning record at home versus the winning record on the road. Does travel distance have anything to do with the road win/loss record? Does crossing more time zones have anything to do with the win/loss record?

_____

_____

_____

7. Does your favorite team play two games in a row on the opposite coast? Do they stay on that coast all week? Why or why not?

_____

_____

_____

From *Football Math: Touchdown Activities and Projects for Grades 4–8* published by Good Year Books. Copyright © 1995, 2005 Jack Coffland and David A. Coffland.

Miles between All NFC Teams and All AFC Cities

Team	Baltimore Ravens	Buffalo Bills	Cincinnati Bengals	Cleveland Browns	Denver Broncos	Houston Texans	Indianapolis Colts	Jacksonville Jaguars	Kansas City Chiefs	Miami Dolphins	New England Patriots	New York Jets	Oakland Raiders	Pittsburgh Steelers	San Diego Chargers	Tennessee Titans
Arizona Cardinals	2,000	1,910	1,560	1,715	594	1,014	1,479	1,797	1,044	1,951	2,295	2,147	643	1,816	313	1,443
Atlanta Falcons	571	723	380	556	1,205	689	448	296	689	573	955	768	2,128	527	1,892	212
Carolina Panthers	364	542	334	435	1,359	927	433	350	804	654	721	534	2,289	364	2,077	340
Chicago Bears	605	478	266	300	892	946	170	874	398	1,183	868	740	1,844	427	1,724	397
Dallas Cowboys	1,210	1,220	812	1,009	648	243	765	911	466	1,104	1,565	1,394	1,472	1,074	1,174	616
Detroit Lions	399	253	230	93	1,119	1,087	223	845	615	1,135	642	518	2,076	227	1,948	472
Green Bay Packers	687	488	339	336	881	1,007	249	1,050	422	1,262	880	766	1,827	468	1,724	582
Minnesota Vikings	947	748	613	623	685	1,074	513	1,207	410	1,505	1,132	1,039	1,562	756	1,534	703
New Orleans Saints	999	1,094	693	901	1,067	307	708	507	685	656	1,363	1,176	1,912	909	1,605	469
New York Giants	173	298	592	441	1,631	1,428	667	846	1,107	1,082	188	1	2,559	331	2,448	761
Philadelphia Eagles	87	275	509	377	1,558	1,334	592	764	1,031	1,005	281	94	2,507	266	2,345	681
San Francisco 49ers	2,543	2,314	2,037	2,147	964	1,655	1,943	2,377	1,508	2,574	2,705	2,588	11	2,267	443	1,960
Seattle Seahawks	2,329	2,114	1,970	2,025	1,017	1,889	1,866	2,461	1,501	2,730	2,488	2,404	678	2,135	1,062	1,972
St. Louis Rams	734	667	314	497	792	678	237	763	234	1,064	1,040	878	1,728	563	1,556	257
Tampa Bay Bucs	851	1,059	778	929	1,516	785	852	162	1,048	183	1,189	1,009	2,388	865	2,093	619
Washington Redskins	34	293	413	321	1,480	1,219	504	656	942	910	399	213	2,427	188	2,277	567

Miles from One NFC Team to Other NFC Cities

Team	Arizona Cardinals	Atlanta Falcons	Carolina Panthers	Chicago Bears	Dallas Cowboys	Detroit Lions	Green Bay Packers	Minnesota Vikings	New Orleans Saints	New York Giants	Philadelphia Eagles	San Francisco 49ers	Seattle Seahawks	St. Louis Rams	Tampa Bay Bucs	Washington Redskins
Arizona Cardinals	*	1,579	1,780	1,434	861	1,656	1,440	1,275	1,298	2,147	2,055	652	1,112	1,265	1,786	483
Atlanta Falcons		*	222	615	730	604	694	932	412	768	666	2,142	2,179	469	405	532
Carolina Panthers			*	589	929	508	747	948	650	534	448	2,299	2,283	572	510	330
Chicago Bears				*	807	226	81	350	834	740	678	1,847	1,783	262	1,018	1,038
Dallas Cowboys					*	984	855	872	449	1,394	1,302	1,470	1,680	544	932	359
Detroit Lions						*	250	532	919	518	463	2,071	1,931	457	991	571
Green Bay Packers							*	289	0	766	690	1,826	1,646	423	1,099	686
Minnesota Vikings								*	1,056	1,039	992	1,585	1,382	467	1,333	838
New Orleans Saints									*	1,176	1,083	1,918	2,098	599	488	956
New York Giants										*	94	2,577	2,404	878	1,009	103
Philadelphia Eagles											*	2,471	2,374	810	928	120
San Francisco 49ers												*	684	1,739	2,390	2,113
Seattle Seahawks													*	1,717	2,526	2,324
St. Louis Rams														*	863	711
Tampa Bay Bucs															*	818
Washington Redskins																*

Miles from One AFC Team to Other AFC Cities

Team	Baltimore Ravens	Buffalo Bills	Cincinnati Bengals	Cleveland Browns	Denver Broncos	Houston Texans	Indianapolis Colts	Jacksonville Jaguars	Kansas City Chiefs	Miami Dolphins	New England Patriots	New York Jets	Oakland Raiders	Pittsburgh Steelers	San Diego Chargers	Tennessee Titans
Baltimore Ravens	*	275	420	304	1,507	1,251	508	689	959	958	360	173	2,442	194	2,291	595
Buffalo Bills		*	417	176	1,367	1,299	455	892	868	1,174	392	298	2,289	197	1,166	630
Cincinnati Bengals			*	211	1,071	884	105	636	532	932	755	592	2,033	264	1,864	239
Cleveland Browns				*	1,189	1,092	245	780	671	1,066	580	441	2,167	133	2,012	460
Denver Broncos					*	880	976	1,475	543	1,697	1,758	1,631	937	1,306	849	1,023
Houston Texans						*	864	825	666	940	1,609	1,428	1,633	1,130	1,315	665
Indianapolis Colts							*	715	441	1,014	821	667	1,936	337	1,780	259
Jacksonville Jaguars								*	961	317	1,024	846	2,368	714	2,093	508
Kansas City Chiefs									*	1,230	1,251	1,107	1,494	777	1,341	475
Miami Dolphins										*	1,251	1,082	2,579	987	2,256	815
New England Patriots											*	188	2,685	491	2,589	943
New York Jets												*	2,559	331	2,448	761
Oakland Raiders													*	2,253	452	1,950
Pittsburgh Steelers														*	2,117	473
San Diego Chargers															*	1,740
Tennessee Titans																*

From *Football Math: Touchdown Activities and Projects for Grades 4–8* published by Good Year Books. Copyright © 1995, 2005 Jack Coffland and David A. Coffland.

Collecting Trivia Information

Many people are interested in sports trivia. Trivia questions deal with interesting, isolated facts that stump most people, but not the "trivia buffs." You can have a good time making up questions that your friends will have a tough time answering.

The following four pages include trivia questions for both college and professional teams, with both modern and historical questions.

1. Can you answer the questions? Can your mom or dad?

2. Can you stump your friends with these questions?

3. Can you make up more questions on your own?

Where can you get the information to make up trivia questions?

1. Try your library. They will have sports information books.

2. Try a bookstore in your town. They will have a sports section with sports reference books in them. You may also find record books that list records from different sports.

3. Keep a record of trivia information from your newspaper for an entire season. Then write questions about what you have collected.

Historical NFL Trivia

1. Only two of the original fourteen teams who founded the NFL in 1920 have an unbroken history to the present. However, both teams have changed cities and one has changed its nickname. What two modern-day teams trace their origins to the 1920 season? How many years have they been in the NFL?

   _____

   _____

2. What is the team with the longest unbroken record of playing in one city with the same nickname? *Hint:* Their first year in the NFL was the 1921 season, and their record for that season was three wins, two losses, and one tie. How many years has this team been in the league?

   _____

   _____

3. Teams in the early years of the NFL did not play a regular schedule nor the same number of games. Several oddities occurred because of this fact in the first ten seasons of the NFL. For example:

a. What team played the fewest of games in the NFL? Did they win any games? How many games did they play?

   _____

   _____

b. What team played in both 1920 and 1921 before dropping out of the league, but played only a total of three games over the two seasons and lost them all?

   _____

c. What team played the most regular-season games in the NFL? How many regular-season games did they play? How many years ago? What was their final record for that year?

   _____

   _____

d. What was the original name of the NFL?

   _____

e. What 2 teams played the first game between two official NFL teams? What was the score? Who scored the first TD? What was the date?

   _____

   _____

4. Jim Thorpe, the famous Native American football player of the 1920s, was both the league president and an active player in 1920. For what team did he play?

   _____

From *Football Math: Touchdown Activities and Projects for Grades 4–8* published by Good Year Books. Copyright © 1995, 2005 Jack Coffland and David A. Coffland.

1. Everyone knew that Dan Marino set the record for TD passes in one season, as he threw for 48 touchdowns in 1984. But what two players held the record for touchdown passes in a season before Marino? How many fewer points did their teams score than the Dolphins scored on Marino's passes?

_____

_____

What player broke Marino's season TD record in the 2004 season?

_____

2. Nine of the ten most watched TV programs (those with the largest number of people watching their TV sets) were Super Bowls. What was the one TV program which was not a Super Bowl that broke into this top 10 list?

_____

_____

3. Which Super Bowl had the largest crowd in attendance? Where was it played? Who were the teams? What was the score? Who won? How many people were in the stands?

_____

_____

_____

4. What present AFC team played only one season as an NFC team? There is also a reverse of this question: What present NFC team played one season as an AFC team? What three AFC teams began in the National Football League and switched to the AFC at the time of the merger?

_____

_____

5. Some records are dubious records. For example, who holds the record for the most fumbles in a career in the NFL?

_____

6. Which passer threw the most interceptions in his career? *Hint:* He also holds the record for most interceptions in a single season.

_____

Historical College Trivia

I. The Rose Bowl was the first bowl game played at the end of the college season. Historically, any team could be invited to play. (Now the champions of the Big 10 play the PAC 10 champs.) So what four teams have won games in the Rose, Orange, Sugar, and Cotton bowls?

_____

_____

_____

_____

2. What two college teams have played the most games against each other over the years?

_____

_____

3. What is the longest winning streak in college football history? Who was the team, how many games did they win, and over what seasons did the streak stretch?

_____

_____

_____

4. What was the longest losing streak in college football history? Who was the team, how many games did they lose in a row, and over what seasons did the streak stretch?

_____

_____

_____

5. In 1936, which team won the first "national title" as voted on by the Associated Press writers' poll?

_____

6. During World War II, there was a year when the Rose Bowl could not be played in California due to fears that the stadium might be bombed by the Japanese. The game was moved to the East Coast. What year was the game played, where was it played, and what teams played in it? Who won? By what score?

_____

_____

_____

_____

From Football Math: Touchdown Activities and Projects for Grades 4–8 published by Good Year Books. Copyright © 1995, 2005 Jack Coffland and David A. Coffland.

1. What quarterback attempted the most passes in one game? How many passes did he attempt? How many were completed? How many yards did he gain? For what team? Against who? Did the team win or lose?

_____

_____

_____

_____

_____

2. The mythical college championship is determined by voter polls. This system has been in place since 1936. How many years has it existed? How many times have their been co-champions? (Consider only the major polls—the AP writers' poll and the UPI/CNN coaches' poll.)

_____

_____

3. When was the first time that the number 1 ranked team played the number 2 ranked team?

_____

4. In the 1980s, one college team played the number 1 ranked team seven times and won every one of those games. Who was that team?

_____

5. Division I-AA football is played by colleges who are not as large as teams like Nebraska and UCLA. There is a play-off system to determine the champs of I-AA football. In 1991, '92, and '93 the same two teams played for the championship each year. Who were they? Who won each year?

_____

_____

6. What Ivy League school has won more football games than any other IAA school?

What major (Division I-A) college has won more games than anyone?

_____

_____

Answer Key

Answers are provided for most activities.
For activities that do not appear here, answers vary.

Activities

2–Miami Dolphin Records
1. 317.75, or 318 yards per game
2. 288 points
3. 1,030 pass receptions
4. 5,500 yards of total offense
5. 869 more yards of total offense
6. 4.473, or 4.5 yards per carry

3–Silver and Black Attack
1. 66 more games
2. 612 points scored on Brown's TDs
3. Rice (end of 2003 season) had 449 more catches
4. Allen gained 3,398 yards for Kansas City

4–Debbie's Dogs
1. 1,500 hot dogs
2. 1,200 cups
3. 956 soft drinks
4. 1,848 nachos
5. 1,305 cups
6. 2,388 cups

5–Debbie's Dollars and Sense
1. $765.88
2. $449.28
3. $59.04
4. $348.70
5. $1,308.30
6. $2,931.20

6–Great Teams and Great Players
1. 66 points
2. 20.75 passes per game
3. 48 points
4. 80.25 yards per game
5. 307.8 yards per game

7–Deee-fense
1. 15 fumbles
2. 30.4%
3. 30%
4. 140 tackles

9–Beasts of the East
1. 1,497 yards per team or 1,497.0 yards
2. 3,962.8 yards per team
3. 2,431.8 more passing than running yards

Challenge problems
You must find the new statistics and compare them to answers for Problem 1 (rushing) and Problem 2 (passing).

10–Bobcats Victory
1. 177 yards
2. 11.4 yards/catch
3. 8.6 yards/rush
4. 40.3 yards/punt
5. 89.4 yards

11–Why Is It Called Football?
1. 456 points
2. 117 points on field goals
3. 22 field goals
4. 28 more points
5. 156 extra points

Challenge problem
29 goes into the "Extra Points" blank.

12–They Get a Kick Out of This
1. 27 field goals (Your answer of 26.97 needs to be rounded off to the nearest whole number, because you can't make part of a field goal!)
2. 98.7% of the PATs were made by the four kickers.
3. 34 field goals attempted
4. 84.7% of the field goals attempted were made
5. 66.7% of Mare's 50+ yard field goals were made

13–AFC Rushing Leaders
1. 421 more yards
2. 2,622 total yards. They came on 602 carries.

These running backs probably will be missed by their old teams, because that many carries and yards gained are always hard to replace.

3. Priest Holmes's longest run was 4 yards longer than that of Eddie George.
4. Williams had 154 more carries than Davis.
5. He scored 162 points.
6. He didn't rush for more yards than James and George combined. Together, they rushed for 224 more yards than Lewis. But Lewis did gain more 4 yards than the combined total for Eddie George and D. Davis.

15–NFC Rushing Leaders
1. The two combined to rush for 3,524 yards.
2. Alexander scored 84 points for his team.
3. Green had 154 more carries than Barlow.
4. Green's longest run was 71 yards longer than Barber's.
5. Green averaged 22.19 carries per game. McAllister averaged 21.94 carries per game.

From *Football Math: Touchdown Activities and Projects for Grades 4–8* published by Good Year Books. Copyright © 1995, 2005 Jack Coffland and David A. Coffland.

Davis averaged 19.98 carries per game. Alexander averaged 20.38 carries per game.

Challenge problem

Green gained approximately 1 more yard every time he carried the ball. While that is not a large figure, it probably resulted in more first downs, more long runs, and so on.

17–Home Field Advantage

1. Denver's home field winning percentage was 75%, but it won less than half of its road games, or 48%.
2. Miami won 70% of its home games, but it won only half of its road games, or 50%.
3. Green Bay won 85% of its home games. You should know they won 50% of their road games, as their record is the same as Miami's road record.
4. Tampa Bay won 65% of its home games, and only 43% of its road games. They had the worst road record of any of the four teams.
5. Combined Home Record: Wins: 236; Losses: 84; Winning %: 74
   Combined Road Record: Wins: 152; Losses: 168; Winning %: 48
   Total Record: Wins: 388; Losses: 252; Winning %: 61

19–All-time Winning Coaches

1. Shula had 23 more total wins and 10 more regular-season wins.

2. Shula had 19 post-season wins and 17 post-season losses. Halas had 6 post-season wins and 3 post-season losses.

Why the big difference? More post-season games were played during Shula's NFL coaching career. George Halas coached many years before the extended play-off season and the Super Bowl came to the NFL.

2. 23 to tie, 24 to pass him on the Total Wins list
3. 803 games
4. 7 games/year
5. Shula has 146 more total wins than Reeves.

Challenge problem

Shula coached games after the NFL tie breaking system started; Halas did not. When Halas coached, games could end in ties at the end of the 4th quarter.

21–Home Cooking— Different Flavors

1.

	Home Winning %	Away Winning %
Eagles	66%	78%
Vikings	69%	28%
Colts	59%	59%
Chiefs	69%	38%

2.

	Home Winning %	Away Winning %	Diff.
Eagles	66%	78%	–12%
Vikings	69%	28%	41%
Colts	59%	59%	Same
Chiefs	69%	38%	31%

3. Minnesota Vikings
4. Indianapolis Colts
5. Minnesota Vikings and Kansas City Chiefs
6. Minnesota Vikings

22–Comparing the "Best of the Best" Coaches

	Average Wins Per Season	% of Games Won
Shula	9.94	67%
Halas	7.95	64%
Landry	8.62	60%
Lambeau	6.85	59%
Brown	8.52	65%

Challenge problems

Modern teams play more games, so it is probably not fair to compare the "total wins per season."
Figuring the percentage of games won over games played is probably a better way to compare coaching records.

23–Monsters of the Midway

1. 132.3 yards per game
2. 4,400 yards rushing in a single season
3. 1,367 total passing yards
4. 729 points
5. Game = 36 points
   Season = 132 points

25–St. Louis Rams Records

1. Dickerson averaged 1,449 yards per season
2. Bruce averaged 111.3 yards per game
3a. Hirsch averaged 118.75 yards per game
3b. Hirsch would gain a total of 1,900 yards with that average in a 16-game season.
4. Answers vary by year. By 2004, the record had stood for 53 years
5. 5,376 total yards in a season

Challenge problem

If Tom Fears caught 18 passes a game for all 12

Answer Key

games in his season, he would still break the record. He would catch 216 passes which is more than twice the present record. Can you find the record for "Most passes caught in one NFL season"?

27–Cleveland Brown Heroes
1. 9.67 wins per season
2. 13,506 yards
3. 4,414 more yards
4. Brown. He averaged 1,368 yards per season. Payton averaged 1,286 yards per season.

Challenge problem
Groza kicked 647 extra points.

28–Cincinnati Bengals History
1. 36 points
2. 115 games
3. 115 wins for other teams
4. 17 more games
5. 42 more losses
6. 65 wins

29–Indianapolis Colts History
1. 2,339 yards gained passing each season
2. 12.36 average yards per completion
3. James averaged 4.42 yards per carry.
4. Harrison averaged 107.63 yards per game.
5. Berry averaged 104.83 yards per game.

30–Lincoln's Big Win
1. 604 yards
2. 32 yards
3. 88 yards
4. 420 pushups
5. 76 players
6. 380 cups

31–Halftime Munchies
1. $8.00
2. $10.00
3. 51 points
4. $21.00
5. $1.50

32–All-purpose Yards Leaders I
1. 231 yards
2. 237 yards
3. 312 yards

33–College Quarterback Ratings
74.51 rating

35–Forward and Backward
1. 353 total yards
2. 434 net yards
3. 382 net yards
4. 302 total yards
5. 631 net yards
6. –59 yards

37–Football Fashions
1. $1,225.00
   $ 980.00
   $ 630.00
   $ 280.00
   $ 210.00
   $1,575.00
   $1,620.00
   $ 720.00
   $2,340.00
   $ 175.00
   $ 184.00
2. $ 540.00
3. $ 140.00
4. $ 138.00
5. 28
6. $ 558.00

39–New York Jets records
1. 558 PATs
2. 18 points
3. 444 points
4. 252 points
5. 1,297 yards per season

40–Buffalo Bills Records
1. 37,405 yards
2. 1,647 yards
3. 14.6 yards per catch
4. 99 points
5. 36,876 yards

Challenge problem
Dan Marino 61,361 yards
As of the start of 2004, Bledsoe was 24,485 yards behind.

42–Dallas Cowboys Super Bowls
1. 25 years
2. 221 points
3. 132 points
4. Dallas, 27.6 points per game; Opponents, 16.5 points per game
5. $181,000
6. $2,438,000

43–Detroit Lions Stars
1. 6,793 yards
2. 8,476 yards
3. 63 points
4. 405 points

44–Kansas City Chiefs History
1. 82 minutes, 40 seconds
2. 77 minutes, 54 seconds
3. 6 TDs
4. 35 points

45–Crazy College Capers
1. 25
2. 54 yards
3. 25–20; No PAT was attempted.

46–Pre-game Pep
1. $106.00
2. 53 banners
3. 8-1/6 yards
4. 12 strips
5. 5-5/12 yards

48–Quarterback Stats

1. 77 completions
2. 42%
3. 123 attempts
4. 46%
5. 105 attempts
6. 50 completions

49–Frantic Fans

1. Solve this problem using a "guess and check" strategy. Guess two numbers and check to see if the directions work. Jerry must begin with seven cards and Barbara must begin with five cards. If this is the case, then when Jerry tries to give one to Barbara, they would both have six cards. But when Barbara refuses that and then gives one card to Jerry, Jerry has eight cards and Barbara is reduced to four. Thus Jerry would have twice as many as Barbara.

2. The key to this problem is the phrase "each club member paid the same amount." That means something must divide into $2.89 evenly. When we test the possible factors of 289, we find that the number can be divided evenly in only one way: 289 ÷ 17 = 17. Therefore, 17 club members each paid 17 cents in dues. Again, there is only one way to use 5 coins to make 17 cents: 3 nickels and 2 pennies. Therefore, 17 club members paid 3 nickels each, so the club received 51 nickels.

3. There are six combinations of three whole numbers that multiply together to equal 24. They are:

    1, 1, 24
    1, 2, 12
    1, 3, 8
    1, 4, 6
    2, 2, 6
    2, 3, 4

    Only the fourth group, 1, 4, and 6, add up to be 11. Therefore Kyle and Conner's favorite players have 1, 4, and 6 years of experience.

50—How Big is the Team?

1. The correct answer is 26 players. This solution is shown in the Venn Diagram:

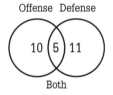

Offense Defense

Both

A common mistake is to add 15 and 16 to get a "solution" of 31. This method does not take into account that 5 players play on both squads. These players are counted twice in the mistaken procedure. A correct procedure is to note that 5 offensive players is 15, then 10 (15 – 5) players are offensive specialists. By similar logic, 11 (16 – 5) players are defensive specialists. Adding the number of offense specialists, the number of defensive specialists, and the number of two-way players gives the correct answer of 26 total players.

A second method of calculating the correct answer is to add the number of offensive players (15) to the number of defensive players (16) and then subtract the number of players (5) that play on both squads. This gives the equation 15 + 16 – 5 = 26, which is the correct answer.

2. There are 76 players on the Cougar football team. This solution is shown in the Venn Diagram:

Defense Offense

Special Teams

There are 3 players that play on all 3 squads. This number must be used to determine each of the other numbers. Because there are 11 players who play offense and defense, but 3 play on all the squads, 8 players play offense and defense only. By similar reasoning, 7 players play defense and special teams only and 3 players play on offense and special teams only. Because there are 42 total defensive players, and 18 (8 + 3 + 7) also participate on other squads, 24 players must play defense only. Similar reasoning results in 20 offensive specialists and 11 special-teams-only players. The results of this reasoning are shown in the Venn Diagram above.

Answer Key

(Encourage students to draw diagrams as one method of solving problems.) Adding each of these numbers gives the total of 76 players on the Cougar football team.

51–Rapid Runners

1. This problem requires several steps to solve it. One possible solution is:
   a. Find the average gain that Cesar has for each game:

$387 ÷ 3 = 129$ yards per game

   b. Then, multiply that average by 8 to see what the total projected yards would be for the season, if that pace is continued.

$129 × 8 = 1,032$ yards per season

The answer, then, is that if Cesar holds this average for the remaining five games, he will gain 1,032 yards during the season. That would be excellent for an eight-game season.

2. One of the numbers needed to answer this problem is not given; we must find it. How many yards did Jorge improve between the two years? To find this, we subtract:

$1,093 – 758 = 335$ more yards rushing in his senior year

This is the improvement. Therefore, we must work the percentage problem with this number compared to the original, or the junior year, yardage. The ratio is set up as:

$$\frac{\text{Part}}{\text{Total}} : \frac{\text{Part}}{\text{Total}} \text{ or } \frac{335}{\text{Answer}} : \frac{758}{100}$$

After cross-multiplying, we obtain 33,500/758, which yields a percentage of 44.20%. This is Jorge's percentage improvement.

52–Winner's Edge

1. This is an exercise in following the directions in the best order to get a solution. The limits — b and e—reduce the possible answers quickest. The answer must be between 82 and 98. Using a, it must be an odd number, 83 to 97. Using d, it must be an odd number that is a multiple of 3; this reduces our possible answers to 87 and 93. When we examine c, we find that only 93 gives an even number when you sum the digits. Therefore, the answer is 93.

2. Nineteen points could be scored in a variety of ways, shown as follows. (For ease of showing all answers, the words are omitted from the answer. Students might be asked to write how the score is made using football terms.)

$7 + 7 + 3 + 2 = 19$
$6 + 3 + 3 + 3 + 2 + 2 = 19$
$7 + 6 + 3 + 3 = 19$
$3 + 3 + 3 + 3 + 3 + 2 + 2 = 19$
$7 + 3 + 3 + 3 + 3 = 19$

$3 + 3 + 3 + 2 + 2 + 2 + 2 + 2 = 19$
$6 + 6 + 3 + 2 + 2 = 19$
$3 + 2 + 2 + 2 + 2 + 2 + 2 + 2 + 2 = 19$
$7 + 6 + 6 = 19$
$8 + 8 + 3 = 19$
$8 + 7 + 2 + 2 = 19$
$8 + 6 + 3 + 2 = 19$

14 points could be scored as follows:

$7 + 7 = 14$
$7 + 3 + 2 + 2 = 14$
$6 + 6 + 2 = 14$
$3 + 3 + 2 + 2 + 2 + 2 = 14$
$8 + 6 = 14$
$8 + 3 + 3 = 14$
$8 + 2 + 2 + 2 = 14$

Discussion: Which are most possible?
19 = Three touchdowns, but only one extra point. After missing the second, they went for two on the third and didn't make it.
19 = Two touchdowns, two 2-point conversions, and a field goal
14 = Two touchdowns and two PATs (by far the easiest)

53–Defensive Delights

1. One way to solve this problem would be to make an ordered table. The table would have to show that there were twice as many assisted tackles as solo tackles. Therefore, the table would be:

Solo Tackles	Assisted Tackles	Points
1	2	4
2	4	8
3	6	12
4	8	16

From Football Math: Touchdown Activities and Projects for Grades 4–8 published by Good Year Books. Copyright © 1995, 2005 Jack Coffland and David A. Coffland.

Because we have now reached 16 points with twice as many assisted tackles, we know we have the answer and the table is done.

2. Tank must wear a number between 60 and 79. That is given by the hint. One assumption that can be made is that Tank's number must be in the 70s, because the first digit is a prime number—6 is not prime; 7 is. Because the first digit is a prime number and both digits add up to be that prime number, only 70, 74, and 76 are possible answers. But only 70 gives a 7 for an answer. Therefore, Tank wears the number 70.

54–Scoring Schemes I

1. This problem has an algebraic solution. The equation should be made with an x equaling the number of points scored on place kicks and 4x being the number of points scored on field goals. Therefore:

$$x + 4x = 135$$
$$5x = 135$$
$$x = 27$$
$$4x = 108$$

So, Karl the Kicker scored 27 points on PATs and 108 points on field goals. This means he made 27 PATs and 108 ÷ 3 field goals, or 36 field goals.

Students in grades 4–8 do not know algebra; they will have to invent the solution. They

may use several different strategies, "guessing and checking" or making a systematic list to get the solution.

2. One way to solve this is to make a systematic list. Because the Hokies scored seven times, they will have between 7 TDs/0 field goals and 0 TDs/7 field goals. Therefore:

TDs	FGs	Total Points
7	0	49
6	1	45
5	2	41
4	3	37
3	4	33 (the score)
2	5	29
1	6	25
0	7	21

Therefore, the Hokies scored 3 touchdowns and 4 field goals.

55–Helpful Heroes I

1. Because Toady Tom is so fussy, Bill will have to do the following:

Step 1: Fill up the 3-cup container. Pour that water into the 5-cup container.

Step 2: Then fill up the 3-cup container again and pour as much water into the 5-cup container as you can. That will be 2 cups, leaving 1 cup in the 3-cup container.

Step 3: Empty the 5-cup container. Then pour in the 1 cup of water left in the 3-cup container.

Step 4: Then fill up the 3-cup container again and pour into the 5-cup container.

The 5-cup container will now hold the 1 cup + 3 cups, which equals 4 cups of water. Toady Tom can now have his drink.

2. The simplest way to answer this question is to make a list of the number of pushups the Tiger does after each TD. The problem is simplified by the fact that Clemson only scored touchdowns. The problem solver does not have to consider other scoring plays. First, 84÷7 = 12 TDs.

After Clemson's
1st TD: 7 pushups
2nd TD: 14 pushups
3rd TD: 21 pushups
4th TD: 28 pushups
5th TD: 35 pushups
6th TD: 42 pushups
7th TD: 49 pushups
8th TD: 56 pushups
9th TD: 63 pushups
10th TD: 70 pushups
11th TD: 77 pushups
12th TD: 84 pushups
Total: 546 pushups

Mathematician Karl Frederick Gauss developed a formula for finding the sum of the first n counting numbers; the formula can be modified to obtain the answer to this problem. Gauss's formula is:

$$S = n \times (n + 1)$$, where n is the number of numbers being added.

Answer Key

In this problem, n = 12 because Clemson scored 12 touchdowns and we are finding the sum of the first 12 counting numbers. This formula gives an answer of 78. But we are working with touchdowns, so we must multiply this answer by 7. This provides us with 78 × 7 = 546. This logic is a simplified version of what the mathematician would actually do; but it is a good example of how a student may modify a formula to solve a problem.

56–College and Professional Game Records

All percentages are rounded to the nearest tenth.
1. 72.7%
2. 83.0%
3. 77.4%
4. 64.3%
5. 66.3%

57–Scoring Schemes II

1. Because, in this problem, Seattle scored 10 points less than half as many as New England, we have to work backward. New England scored 186 points. Half of that figure is 93 points. Ten less is 83 points. Therefore, Seattle would have scored 83 points in this situation.

   Actually, Seattle scored 140 points. The next lowest team in the NFL was New England, who scored 205 points. Seattle did have a "bad year." In fact, Steve Christie, the kicker for Buffalo, scored 115 by himself. He almost out scored Seattle single-handedly.

2a. The algebraic formula and computation for this problem would be:

   $1 \div 2x + 1 \div 4x + 1 \div 6x + 3 = x$. Find a common denominator.

   $6 \div 12x + 3 \div 12x + 2 \div 12x + 3 = x$
   $11 \div 12x + 3 = x$
   $3 = x - 11 \div 12x$
   $3 = 1 \div 12x$
   $36 = x$

   Sam has played in 36 games.

   b. and c. He has scored 3 TDs in 18 games, 2 TDs in 9 games, and 1 TD in 6 games. He did not score in 3 games: 18 + 9 + 6 + 3 = 36. He scored, then, 18(3) + 9(2) + 6(1) = 78 TDs, which equals 468 points in the 36 games.

58–Helpful Heroes II

1. One way to find the average speed is to divide the total miles by the total time. Because the distance for the first part of the trip is not given, it must be found. The time, 40 minutes, is 2/3 of an hour. At 30 miles per hour times 2/3 of an hour, the bus went 20 miles in the snowstorm. Add this to the 105 miles on the freeway, for a total distance of 125 miles.

   It is also necessary to find the total time. Because the time for the second part of the trip is not given, it must be found. Divide the 105 miles on the freeway by 75 miles per hour to get 1-1/2 hours. Add the 1-1/2 hours to the 2/3 of an hour from the first part of the trip to get a total time of 2-1/6 hours of travel. Finally, divide 125 miles by 2-1/6 hours to get an average speed of 57.7 miles per hour for the whole trip. Because the total travel time for the trip was 2 1/6 hours and the team allowed 2 hours for travel, the team bus arrived 1/6 of an hour late. Because 1/6 of an hour is 10 minutes, you could also say they were 10 minutes late.

2. This problem requires simultaneous equations, a topic that is not taught until first- or second-year algebra. That procedure is not available to students in grades 4–8. They must use another procedure. Making a systematic list will get the answer. The problem with this solution is that there are 31 different whole-number combinations to examine, but there is no guarantee the answer will not include decimals. That possibility increases the list to a potentially infinite series of combinations. Using a "guess and check" strategy, along with some logical thinking, can reduce the number of combinations to be checked. Consider:

Answer Key

From *Football Math: Touchdown Activities and Projects for Grades 4–8* published by Good Year Books. Copyright © 1995, 2005 Jack Coffland and David A. Coffland.

First: Let's "guess" that they made equal amounts. What would happen if Fred and Carrie both made 15 pounds of caramel corn?

Carrie's 15 pounds is 70% popcorn. This is 10.5 pounds of popcorn.

Fred's 15 pounds is 60% popcorn. This is 9 pounds of popcorn.

Together, they would have used 19.5 pounds of popcorn.

This is more than the 19 pounds of popcorn available. Logic: The 19.5 pounds of popcorn is too much. Because Carrie uses more popcorn in her mix, our next guess should reduce Carrie's amount:

Second: Let's "guess" that Carrie made 12 pounds and Fred made 18 pounds.

Carrie's 12 pounds is 70% popcorn. This is 8.4 pounds of popcorn.

Fred's 18 pounds is 60% popcorn. This is 10.8 pounds of popcorn.

Together, they would have used 19.2 pounds of popcorn.

This is still more than the 19 pounds of popcorn available. Logic: The answer is still too large. Reduce Carrie's amount more.

Third: Let's "guess" that Carrie makes 9 pounds and

Fred 21 pounds. Using the same procedure, we see that this guess uses only 18.9 pounds of popcorn. We are now under the 19 pounds of popcorn. We've gone too far. Using the same procedure, we see that this is the correct answer.

Fourth: Now, let's guess that Carrie makes 10 pounds of caramel corn. Can you do the math from there?

Mini-projects and long-term projects: Answers will vary.

Collecting Trivia Information
106–Historical NFL trivia

1. Chicago Bears. Began as the Decatur Staleys in 1920, moved to Chicago in 1921, and played as the Chicago Staleys. Became the Chicago Bears in 1922. George Halas bought the team in 1921.

   Phoenix Cardinals. Started as the Chicago Cardinals. Played in Chicago from 1920 to 1959. Moved to St. Louis and played there from 1960 to 1987. They moved to Phoenix in 1988 and have played there ever since.

2. The Green Bay Packers began play in the NFL in 1921. They have not moved, nor have they changed their name since that time.

3a. The Tonawanda Kardex. They played 1 game in the 1921 season and lost. They concluded their NFL career

with a record of 0 winds and 1 loss.

3b. The Muncie Flyers. They played and lost 1 game in the 1920 season. They played and lost 2 games in the 1921 season. Their combined NFL record, therefore, is 0 wins and 3 losses.

3c. The Frankford Yellow Jackets. In the 1925 season, they played 20 regular-season games. Their record was 13 wins and 7 losses.

3d. The first meeting in 1920 selected the name "The American Professional Football Conference." At the second organizational meeting, the name was changed to "The American Professional Football Association." The first games in 1920 were played under the APFA banner.

3e. The first game between two APFA teams was played on October 3, 1920. The Dayton Triangles hosted the Columbus Panhandles. Dayton won, 14–0. The first touchdown was scored by Lou Partlow of Dayton. (Later that same day, the Rock Island Independents defeated the Muncie Flyers 45–0. This game ended Muncie's season that year.)

4. Jim Thorpe played for the Canton Bulldogs. Canton is now the home of the Professional Football Hall of Fame.

Answer Key

107–Modern NFL Trivia

1. Y. A. Tittle and George Blanda each threw 36 TD passes in a season. Thus, they were responsible for 216 points. (This figures 6 points for the TD, as the kicker gets the extra point.)

2. The last episode of M*A*S*H was a 2-hour movie shown on February 28, 1983. This movie ranked number 4 on the all-time list. All the other top 10 shows were Super Bowls. Officially, Super Bowls 27, 20, 21, 26, 19, 22, 25, 23, and 16—in that order.

3. Super Bowl 14, played in the Rose Bowl in Pasadena California. There were 103,985 people in the stands. The Pittsburgh Steelers defeated the Los Angeles Rams by a score of 31–19.

4. As of 2004:

   Two teams have switched leagues twice:
   Seattle Seahawks
      1976 NFC
      1977 to 2002 AFC
      2003 Back to NFC
   Tampa Bay Buccaneers
      1976 AFC
      1977 To NFC

   In addition, three teams switched leagues in 1970 when the NFC/AFC alignment was formed: the Cleveland Browns, the Pittsburgh Steelers, and the Baltimore (Now Indianapolis) Colts.

Note: Cleveland has another interesting aspect to its history, occurring in the late 1990s. The players and owner left to form the Baltimore Ravens, but the Browns franchise and all of its history remained in Cleveland. The team was out of the NFL for a short period of time, and then it reformed as an expansion franchise.

5. Warren Moon had 161 fumbles in his career.

108–Historical College Trivia

1. Alabama, Georgia, Georgia Tech, and Notre Dame. All of these teams played in the Rose Bowl before 1947, when the present Big 10 and PAC 10 playoff started in the Rose Bowl.

2. Lafayette and Lehigh. They have played each other more than 140 times. No other rivalry is close to this number. Can you find the exact number for this year?

3. Oklahoma won 47 games in a row over seasons spanning 1953–57

4. The Columbia Lions lost 44 games in a row over seasons spanning 1983–88.

5. The Minnesota Gophers

6. In 1942, Oregon State played Duke for the Rose Bowl game, but the game was held at Duke Stadium in Durham, North Carolina.

Oregon State won the game, 20–16.

109–Modern College Trivia

1. Paul Gray of Hanover College in Indiana threw 92 passes in one game. He completed 41 of them for 630 yards of total offense and 7 TDs, but his team still lost to Georgetown (KY) College, 55–46. This was a NAIA Division II game. He was quoted as saying "I'm numb" after the game.

2. Answers will vary.

3. On October 9, 1943, number 2 Michigan beat number 1 Notre Dame by a score of 35–12.

4. The Miami Hurricanes

5. The teams were the Marshall "Thundering Herd" and the "Youngstown Penguins." Youngstown won in 1991 and 1993; Marshall won in 1992. The scores were:
   1991: Youngstown 25, Marshall 17
   1992: Marshall 31, Youngstown 28
   1993: Youngstown 17, Marshall 5

6. Yale has won 809 games, more than any other 1AA college or university. Michigan has won 813 games, more than any college or university. In fact, Michigan passed Yale as the college with the most wins between 2000 and 2004.